The Handbook of Pesticides

The Handbook of Pesticides

Hazem Shawky Fouda

Delve
PUBLISHING

www.delvepublishing.com

The Handbook of Pesticides

Hazem Shawky Fouda

Delve Publishing
224 Shoreacres Road
Burlington, ON L7L 2H2
Canada
www.delvepublishing.com
Tel: 001-289-291-7705
001-905-616-2116
Fax: 001-289-291-7601
Email: orders@arclereducation.com

ISBN: 978-1-77407-691-0 (Hardcover)

ABOUT THE AUTHOR

Hazem Shawky Fouda has a PhD. In Agriculture Sciences, obtained his PhD. from the Faculty of Agriculture, Alexandria University, 2008, MSc. In Agriculture Sciences from the Faculty of Agriculture, Alexandria University in 2004, Post-Grade Diploma in Cotton, 2001, BSc. in Agriculture Sciences, from the Faculty of Agriculture, Alexandria University, 1997, He is working in Cotton Arbitration & Testing General Organization (CATGO) since 1999 till present. Was working in the International Cotton Training Center (ICTC) – Cotton Arbitration & Testing General Organization (CATGO) from 2000 till 2015, as a Lecturer & Classer's Trainer for Egyptian and foreigner classers, technicians, ginners, spinners & traders in all cotton aspects starting from picking seed cotton till preparing bales for export including other cotton aspects and training them practically on classing cotton. Besides that he was an editor and active member in the Research & Translation Committee, participating in issuing weekly, monthly and annually issues about the international & local cotton market including price trends and direction, recent developments & researches concerning cotton production, protection, harvesting, ginning, fiber testing, spinning & weaving since its foundation in 2000 till 2014 and since 2015 till 2018 he worked as an inspector from 2018 till present works as a consultant..

TABLE OF CONTENTS

LIST OF FIGURES

LIST OF TABLES

ACKNOLEDGEMENT

First and foremost, I feel always indebted to **ALLAH**, the most beneficent and merciful.

Words stand short when they came to express my gratefulness to my late parents **Shawky Fouda** and **Fatma El-Hussainy**, my uncle **Ahmed Fouda**, sister **Riham Fouda**, my wife **Heba Sadek**, my son **Mohammed Fouda**, my daughter **Raghad Fouda,** relatives and friends for their support, encouragement, care and patience and for their love.

DEDICATION

This piece of work is dedicated to the memory of my late parents **Shawky Fouda** and **Fatma El-Hussainy** (may God bless their souls).

PREFACE

Because of the increase in agricultural food production in the world presently, it has become increasingly significant to try and evaluate the usage of pesticides and their impacts on human health and the environment. Better knowledge of pesticides and various factors that are limiting their usage around the world will permit a better understanding and interpretation.

Therefore, this book was compiled with utmost scrutiny of different journals, articles, research, and a wide array of books. The main objective of preparing this book is to actually contribute towards the development of knowledge base in the field of pesticide usage.

Not only can this book be used by a wide range of readers who are directly or indirectly working in pesticide application fields but also by a wide range of readers. These include undergraduate students, postgraduate students, scientists and those working at research institutes, planners, implementing agencies, and also various governmental and non-governmental agencies working in pesticide application fields.

The book has ten chapters, and each chapter discusses about a specific topic under pesticides: chapter one all about introduction to pesticides; chapter two is about pest control; chapter three discusses about classification of pesticides; chapter four discusses about benefits and risks associated with the use of pesticides; chapter five presents information about how pesticides get into the environment; chapter six discusses about children exposed to pesticides; chapter seven offers details about toxic kinetics changes for different types of pesticides; chapter eight offer information on what can be done to prevent pesticide exposure and poisoning; chapter nine discusses about handling pesticides; and lastly, chapter is all about rules and treaties that govern pesticides.

These ten chapters were comprehensively prepared for the readers to have a glimpse of pesticide usage. Great focus and emphasis were given to the simplicity of the content in order to make the book comprehensible and understandable to all readers

CHAPTER 1

Introduction to Pesticides

CONTENTS

1.1 INTRODUCTION

"Pesticide" is a generic term which does not tell much about a particular chemical. In monitoring pesticides, it is important to have a working knowledge of pesticide terms and the ways in which pesticides are classified. Pesticides have been widely used in controlling pests and diseases in agriculture, forestry, veterinary and public health. They also play a major role in maintaining weed-free lawns, sports grounds and in clearing roadsides.

Pest- induced losses due to plant diseases, arthropod and vertebrate pests, as well as weed infestations, are estimated to be as high as 30-35% in the field. Post-harvest losses are estimated at 40-50%. It is estimated that without pesticides, world losses caused by pests in agriculture are in the order of US$ 75,000 million annually or about 35% of potential production. Pesticides have played a major role in reducing deaths due to insect-borne and tick-borne diseases of man such as malaria, trypanosomiasis, typhoid, yellow fever, tick fever and others. For example, after the discovery of DDT, malaria was eradicated for some 619 million people who lived in malarious areas and in India, the number of malaria cases dropped from 100 million between 1933-1935. Vector-borne diseases of livestock have also been controlled. Despite their beneficial effects, pesticides can be detrimental to humans. Misuse, mishandling and failure to use protective clothing while handling them may lead to exposure and poisoning. Pesticides have been shown to cause many adverse effects in exposed individuals (Subramanyam & Hagstrum, 2012).

Figure 1.1: Collection of pesticides.

Source: https://www.maine.gov/dacf/php/pesticides/public/obsolete_pesticide_collection.shtml

1.2 HOW DOES A PESTICIDE WORK

The benefits of pesticides to humankind are well known. In agriculture, for example, it is estimated that without pesticides, world losses caused by pests are in the order of US$ 75,000 million annually or about 35% of potential production. Towards this loss, insects contribute 13.8%, that of fungi is 11.6%, weeds, contribute to 9.5% and other pests account for 0.1%. According to the latter, field losses are estimated at 30-35% and postharvest losses up to 50%. Pesticides have been widely accepted as an essential input in curtailing these losses. In a case-control study of workers exposed to herbicides in agriculture and forestry between 1950 and mid- 1970s, subjects had a 5.3 times increased risk for soft-tissue sarcoma. In another study, the effect of pesticides on a flower growing population in Colombia; they found that there was an increase in abortions, premature birth and congenital malformations for pregnancies occurring after the mothers began floriculture. The difficulty in assessing the balance between the beneficial and detrimental effects of pesticides has led to what is termed as the "pesticide dilemma". Clearly, the balance can be shifted by minimizing the adverse effects through safe usage and maximizing the benefits by better and more efficient application methods. In addition to preventing loss in yield, pesticides assure the producer of high-quality goods (Subramanyam & Hagstrum, 2012). It is argued that pesticide use permits stability in food production. They do so in three ways:

- The increased storage life of foods
- Enhanced global trade in food between regions by reducing pest transfer
- Increased possibility of using substitute food crops from other areas for emergency food relief

1.3 TYPES OF PESTICIDES

They are different types of pesticides which include:

- Bactericides – used for bacteria
- Fungicides – used for fungi.
- Herbicides – used for plants.
- Insecticides – used for insects.
- Larvicides – used for larvae
- Rodenticides – used for rodents (rats and mice)

One way of classifying pesticides is by their main target pest or function. Pesticides have also been classified by the WHO by hazard, which divides pesticides into three different categories based on their toxicity to laboratory animals. Chemical pesticides have been recommended as pest control agents. They are reported to be effective, selective herbicides. Organophosphate insecticides were introduced by 1960 and by 1967 a wide range of both organochlorines and organophosphates were in use. One of the highlights in the history of pesticide usage; as far back as 1908, carbon disulfide was used as a preservative to store maize grain. At that time, Paris green was also in use for control of cutworms on wheat and tomatoes.

Various insecticidal mixtures were tried in 1933 on coffee. A pyrethrum-kerosene extract was found superior in control of antestia bugs. For the control of dodder, a noxious parasitic weed; a mixture of sulfuric acid, iron sulfate or copper sulfate were used. Various copper compounds were already in the market in the 1940s and were being used for the control of potato blight and other fungal diseases. This decline in tonnage is due to decreased usage of copper-based pesticides which were widely used in coffee in the 1980s but are not used due to the decline in the coffee industry. In addition many currently used pesticides are effective at low concentrations. Therefore, the decline in tonnage suggests an increase in the use of active ingredients effective at low concentrations and not a decline in total usage (Subramanyam & Hagstrum, 2012).

1.4 WHAT INGREDIENTS ARE IN A PESTICIDE PRODUCT

Another mode of classification is by chemical makeup. Almost all pesticides are combinations of 2 or more elements bound together by chemical bonds and can, therefore, be grouped by chemical type.

Table 1.1: Classification of pesticides based on the ingredients

Group	Constituents	Examples
Organochlorines	Carbon, Hydrogen, Chlorine	DDT, Lindane
Organophosphates	Phosphoric acid esters	Dimethoate, Parathion
Carbamates	Carbamic acid esters	Carbaryl, Aldicarb
Phenoxyherbicides	Phenoxyaliphatic acids	2,4-D, 2,4,5-T
Botanicals	Natural plant substances	Nicotine, Pyrethrin

Microbial	Microorganisms	*Bacillus thuringiensis*

Source: https://www.researchgate.net/figure/Structural-groups-pesticide-compounds_tbl1_271674089

Worldwide, it is estimated that pesticide trade is worth $25 billion. There are about 600 active ingredients which have been formulated to over 3000 formulations. This number has remained constant over the last twenty years or so. Factors such as vapor pressure, skin permeability, bio-persistence and viscosity are important parameters to be considered. The use of a less toxic active ingredient is the most obvious improvement that can be made. Effective pesticide dosage rates have been declining over the years. There are two reasons for this. The first is a general departure from routine applications towards pest monitoring followed by timed applications. In addition, the efficacy of pesticides has increased with the discovery of new active ingredients which are effective at very low rates. For example, deltamethrin is effective against various insect pests a rate of 50 g/ha active ingredient while chlorsulfuron, and urea-based herbicide is effective against broadleaved weeds and some annual weeds at 15-35 g/ha active ingredients. Pesticides are important in the maintenance of forests. Outbreaks of pests and diseases on forest trees in the past have been controlled with pesticides (Matthews, 2008).

1.5 USE OF PESTICIDES

Worldwide, it is estimated that pesticide trade is worth $25 billion. There are about 600 active ingredients formulated to over 3000 formulations. This number has remained constant over the last twenty years or so. Many of these pesticides are used globally. The difficulty in assessing the balance between the beneficial and detrimental effects of pesticides has led to what is termed as the “pesticide dilemma”. Clearly, the balance can be shifted by minimizing the adverse effects through safe usage and maximizing the benefits by better and more efficient application methods. Despite their beneficial effects, pesticides can be detrimental to humans. Misuse, mishandling and failure to use protective clothing while handling them may lead to exposure and poisoning.

Pesticides have been shown to cause many adverse effects in exposed individuals. In a case- control study of workers exposed to herbicides in agriculture and forestry between 1950 and mid- the 1970s, cases had a 5.3

times increased risk for soft-tissue sarcoma. In another study, the effect of pesticides on a flower growing population, it was found that there was an increase in abortions, premature birth and congenital malformations for pregnancies occurring after the mothers began floriculture. Many other countries enacted pesticide laws after this. Based on reports on exposure to pesticides it was ended that legislation was needed to reduce the incidence of pesticide poisonings. The Food Agricultural Organization of the United Nations published guidelines for legislation covering the registration of pesticides for sale and marketing (Matthews, 2008).

1.5.1 First use of Synthetic Pesticides

In 1910, the Federal Insecticide Act of the United States was enacted for the purpose of protecting consumers against the misrepresentation of products. The need for comprehensive pesticide laws increased with the discovery of the synthetic organochlorine pesticides in the mid-20th century. Laws were needed to address the problems of safety to crops, to the users and to the environment. The following year, the World Health Organization published a review of the existing legislation on pesticides. A conclusion of the review was that the hazards from pesticides are preventable as long as appropriate legislation is actively enforced. The review found that while most countries in Europe and North America already had pesticide legislation, there was a large variation in the definitions of pests and pesticides and registration data requirements.

The FAO subsequently published guidelines on data requirements for registration, good labeling practice and other areas related to the legislation of pesticides. In designing a regulatory procedure, certain criteria must be addressed; the necessary steps in choosing and designing a pesticide law. Basically, the scopes of the pesticides to be covered by the law and the target pests have to be defined. Generally, the wider the scope, the more comprehensive the law is likely to be. The objectives of the regulatory procedure depend on both the agricultural and economic structure and on legislative and political factors. The basic goal is to control pesticides to enhance the safety of users, bystanders, consumers, non-target organisms and the environment. Pesticides are also used in the treatment of leather and textiles (particularly wool) to prevent attack by fungi and insects in some parts of the world (Matthews, 2008).

Figure 1.2: Synthetic pesticides.

Source: https://www.necn.com/news/new-england/Synthetic-Pesticides-Like-Roundup-Banned-in-Portland-Maine-507973621.html

1.5.2 How Consumption of Synthetic Pesticides Increased Worldwide

World production of pesticides has increased considerably in recent years. In 2000, it was estimated that world pesticide usage stood at 15 million tons worth about $50 billion (US) compared to US$ 10-12 billion in 1980. This trend is expected to continue. The world population is projected at 9 billion by the year 2030. World food production must also increase proportionately to cater to the increased demand for food, energy and fiber. Since good land is already in use, there must be significant gains in average yields. The use of pesticides is expected to continue to make a major contribution to this need. In developed countries, herbicides are widely used in clearing lawns, sports grounds and roadsides.

These are the pesticides in current world usage. The Pesticide Index, however, has about 1500 entries since it includes even those pesticides not being manufactured. In total, there are approximately 600 active ingredients, with about 3000 different formulations available for purchase. This number has remained relatively constant for the last twenty years (Matthews, 2008).

Since the 1980's pesticides have been produced by genetic engineering. Also in plant breeding, specific pest resistance is being incorporated into new varieties of crops and domestic animals. The emergence of transgenic plants, for example, is expected to play a wider role in the future of crop protection and especially in reducing the need for high applications of pesticides and therefore will also help in environmental conservation. Pesticides have been widely used to control pests and diseases of crops, animals and in public health programs. Increased agricultural production and other benefits have been attributed to pesticide usage.

1.6 PESTICIDES USED IN DIFFERENT SETTINGS

1.6.1 Agricultural

Pesticides have been used in agriculture for a long time for the productivity of both crops and animals. However, a number of inherent environmental and health risks result from pesticide usage, particularly if these chemicals are mishandled. Excess pesticide exposure (both accidental and occupational) has resulted in a variety of documented adverse health effects. A large number of pesticide poisonings are due to occupational exposure in agricultural workers. However, it has not been quantified the actual rates of exposure. Past studies showed that farmers used pesticides regularly (primarily on coffee) but frequently mishandled them. On large-scale estates a study reported recently that mishandling of pesticides was still common. The study showed what is commonly accepted in many countries in terms of actual pesticide handling. The information obtained from the spraying exposure study is useful in pinpointing which practices lead to exposure during spraying and which body parts are more at risk of contamination. In different countries, there are no set times to be observed by agricultural workers before entering pesticide sprayed fields. The results of the reentry studies will be useful in setting reentry intervals in different countries for dimethoate, omethoate and synthetic pyrethroids as well as setting pre-harvest intervals for these pesticides on (Matthews, 2008).

The great complication in observational explanatory studies is that a researcher has no control over the allocation of exposure status. Thus it can be difficult to separate the effects of agricultural factors from other potentially confounding factors such as poor nutrition, poor medical access, etc. In experiments, random allocation avoids these problems. The agricultural workplace, however, presents special opportunities for worker exposure.

After the introduction of organophosphorous pesticides, it was recognized that it was possible to get sufficiently exposed to the work environment to become intoxicated. Later, Quinby and Femmon (1958) published a review of case reports in which they drew a convincing association between poisoning symptoms experienced by workers and exposures to parathion residues during harvesting. Earlier, Batchelor and Walker (1954) had shown that farmers are exposed to pesticides during mixing, loading and applying pesticides. One of the earliest attempts to measure applicator exposure was carried out. They used cellulose filter pads to collect pesticides falling on the skin during spraying. This method by affixing patches of blotter paper to the skin or to the outside of clothing with tape. The amount of pesticide extracted from the central portion of the pads on a per unit area basis, was then assumed to characterize the exposure of adjacent parts of the body. This method was also used to extrapolate respiratory exposure and has remained relatively unchanged ever since. The converse would also be true and a spill that misses the patch would result in a considerably lower exposure being estimated. To overcome the shortcomings of the patch method used cotton garments that cover the entire body part during the exposure period. Although this method forms part of the protocol for exposure measurement by the World Health Organization, it has a major disadvantage given the difficulty of extracting pesticides from whole items of clothing; especially given the large amounts of solvents required. The use of gloves for the hands, socks for the feet and tops of socks for both ankles and wrists. Other modifications have also been used.

Figure 1.3: Pesticides use in agriculture.

Source: https://www.chemistryworld.com/news/pesticide-use-threatens-human-rights-un-advisers-say/2500546.article

1.6.2 Veterinary

From ancient times man has competed with other species of animals and plants for the available resources. The use of pesticides in the control of pests on crops and animals can lead to residues in foods. These were reported as far back as 1940 when DDT was found in apples. Today, for many crops, maximum residue levels (MRLs) have been calculated to ensure food safety. This is usually done by the manufacturer and it can be for a variety of crops. Based on this, the Theoretical Daily Intake (TDI) can be calculated for each crop for the amount of that crop that can be consumed in a day. This is also done for worst-case scenarios, for example, infants. Pesticides are used in agriculture and for domestic and veterinary purposes.

In other areas where pesticides are widely used, it has been shown that the general population may also be affected by pesticides. For example, residues of pesticides have been detected in the general population both in other parts of the world and in different countries. Since pesticides are widely used in irrigation schemes, the general population, as well as sprayers, is likely to be exposed. To assess the extent to which pesticide contamination occurs in homes and what are the likely sources of this contamination. Analysis of breast milk samples was done to assess contamination of mothers with persistent pesticides. A survey of health status among occupants of farming households was carried out to assess the extent of pesticide-related health effects among occupationally and non-occupationally exposed people in the scheme. Most farmers also rear animals (Soloneski, 2014).

Figure 1.4: Application of pesticides for tick control.

Source: https://apiowinnie.wordpress.com/2016/10/12/the-power-within/

In domestic animals, pesticides are essential in controlling ticks and other insects that are vectors of various diseases such as East Coast Fever, heartwater, redwater, trypanosomiasis and others. In the tropics, parasitic tick-borne diseases are an obstacle in the development of profitable livestock industry. In companion animals, the use of pesticides is also aesthetic, to reduce external parasite burdens especially fleas and ticks as well as the control of internal parasites some of which infect both man and animals. There are a number of other uses of pesticides. One of the issues in assessing pesticide effects on health is their role as carcinogens. Many organochlorine pesticides have been withdrawn or their agricultural uses severely restricted due to their persistence in the environment, damage to endangered species or the potential to cause chronic health problems or reproductive systems.

Pesticide producing countries are partly to blame for this misuse because more often than not they do not inform importing countries of the full dangers of the chemicals. Several incidents of abuse by pesticide companies have been compiled by the Pesticide Action Network (PAN) since its inception. These include:

- Misleading advertisement
- Inadequate labeling and packaging
- Untrained salespeople providing inappropriate and sometimes dangerous advice

Moreover, many third world countries do not have the expertise or resources to ensure that dangerous chemicals are safely used. For example, in the study area majority of the farmers applied pesticides without wearing any form of protective clothing and they changed the clothes used during pesticide application which increases the chances of pesticide absorption through the skin. These practices result in over three million people being poisoned annually by pesticides must then consent to the import of the chemicals (Soloneski, 2014).

1.6.3 Domestic

The use of pesticides for domestic purposes has led to death from poisoning in other parts of the world. For example, in some countries of pure methomyl powder that was left in a baking tin and mistakenly used to prepare a meal made six people severely ill out of which 3 died. Household exposures may occur from air, food, water or from contact surfaces in the home environment. Indoor air may be contaminated with pest control sprays or from pesticides that are volatile. Unwashed garden produce may contain pesticide residues and may be contaminated by poor agricultural practices. In addition, poor storage and handling within the home can lead to contamination of the home environment. Therefore, monitoring of the environment, referred to as "media monitoring" can characterize the patterns of exposure (Soloneski, 2014). According to previous studies, the magnitude of household exposure to pesticides highlighted outbreaks of acute pesticide poisonings. Forty-two percent were related to contamination of food or water by pesticide formulations. This occurred in relation to:

- Use of seed grain as food
- Improperly applied pesticides
- Spillage in storage or transportation
- Accidental substitution of pesticide for an ingredient normally used as food
- Due to indirect emergence of pesticide in the food chain

Studies on household exposure have been carried out by several researchers. The best demonstrated of these are the U.S based market basket surveys. In one such study, a wide range of foods constituting a balanced diet to feed a 16-19-year-old boy was purchased from ordinary grocery stores in cities in different parts of the country. Some samples contained DDT. Later surveys showed a general declining trend in organochlorine residues (except endosulfan) and a slight increase in the organophosphate malathion. Among the various classes of food, vegetables generally have a high surface: volume ratio. It has been predicted that due to this, they would be likely to carry more surface residues than other foods. This has been confirmed in residue studies; with the exception of P-BHC, the levels of all pesticides were less than those reported (Soloneski, 2014).

Figure 1.5: Indoor pesticide application.

Source: https://www.thespruce.com/indoor-use-of-pesticides-safety-2656652

The use of aldrin, dieldrin and heptachlor was restricted in 1986 and all isomers of hexachlorocyclohexane apart from the y isomer, lindane were banned. In the inventory carried out, aldrin, dieldrin, and lindane were found to be in use and at times were stored in homes in the scheme. Dieldrin and lindane were used on vegetables and aldrin around the home to control termites. It is therefore not surprising that these pesticides were detected in milk. However, this group of pesticides is often misused in the developing countries. For example, farmers use using carbosulfan and endosulfan on vegetables in disregard of regulations restricting their use to seeds and cotton respectively. Other banned chemicals are also used within the scheme and pre-harvest intervals for most pesticides are unknown and not observed since most farmers use unlabeled or mislabeled chemicals so they don't actually know what they are using. More than 70% of these farmers store pesticides inside their homes often not securely. Therefore, pesticides may be within reach of children and may also contaminate houses and indoor air (Soloneski, 2014). Household exposure may also originate from undisposed empty pesticide containers which are occasionally used for domestic purposes.

1.6.4 Institutional

Hazard identification is the evaluation of activities that utilize pesticides and the determination of whether these activities pose a potential risk to health and to the environment. The second step, exposure assessment, evaluates the populations that are or maybe, exposed and the levels of exposure expected. The third step is to establish the dose-response relationship between the intensity of exposure and probability of environmental and health effects. The final stage is risk characterization to estimate the probability various outcomes for a given exposure scenario. These are then combined to formulate management strategies to minimize exposures. In many workplaces well-established toxicology and occupational health procedures have been used to minimize exposure to all chemicals over time. Monitoring the health of workers producing or using pesticides has been found useful for the early detection of health effects. In addition, all persons using or handling toxic chemicals should have a medical evaluation appropriate to the exposure. Such medical evaluations should be appropriate to the risk of exposure; on the medical supervision of workers. Many local, state and institutional agencies have published information on the recognition, treatment and management of pesticide exposure. Exposure may be maintained at a safe level through substitution, reduced usage, education, and protection (WHO, 2010).

Regulations can then define the different measures that will help to control exposure and also ensure that they are carried out. However, the adoption of safe use practices is low. For example, in an audit to investigate the extent of adoption, it was found that less than 30% of the trained farmers were adapting to safety training, high cost of protective clothing, lack of protective clothing in local business centers and the discomfort of wearing available protective clothing. The project also trained agricultural extension staff and retailers of pesticides. The results of these programs are difficult to assess in terms of increased safety. In Guatemala, in addition to farmer training, teachers in rural schools were trained and they, in turn, trained 70,000 children over a three-year period. A special program for children was developed called "the Scarecrow" and was used to highlight the dangers of pesticides to children. However, the benefits of this approach appear not to have been assessed. Both farmers and students were trained on the importance of safety in pesticide handling, especially during application. While considerable reduction in exposure is said have taken place, no data are presented.

Many organizations have developed training programs for chemical safety. For example, many countries emphasis on the Environmental Protection Agency (EPA) worker safety program. The United States Agency for International Development has developed a training program for pesticide users in its agricultural development projects. In countries such as Canada, pesticide applicators, operators and vendors are all required to be trained before they are licensed to operate their businesses. In all institutions, it is advisable for applicators must go through a course on the safe handling of pesticides covering safety and health, protection while spraying, first aid, and poisoning. Therefore, the protection of the skin from contact with pesticides during handling and application would lead to significant reduction in exposure (Soloneski, 2014).

CHAPTER 2

Pest Control

CONTENTS

2.1 INTRODUCTION

Pest control is the management of any creature characterized as an animal, insect, an individual or plant that brings about unfavorable effects on human activities. The human response relies upon the extent of the harm implicated and will deter from the consequences, through prevention and management, with an aim to intentionally kill all the nuisance. Pest control measures are included as an important aspect of an incorporated pest management strategy.

A pest is any animal, insect, plant or any other organism which has a harmful effect on humans, their food and (or) their living conditions. A pest is a species that spreads disease, causes devastation or irritation. Various examples of pests are mosquitoes, rodents, fleas, flies, roaches and weeds. It is important to note that not all insects are pests. Many other insects feed on other insects and are thus helpful insects (natural enemies) and not pests.

Pests can be harmful in many ways like biting, destroy food crops, damage property and make lives miserable. Pest control includes various methods of controlling or eliminating pests using both natural strategies and chemical insecticides. Pest control methods play an important role in various areas such as hospitals, nursing homes, warehouses, restaurants, schools, universities, and product storage.

There are various reasons why we need pest control. One of the most important reasons is that pests that are uncontrolled may spread diseases and cause contamination in and outside of homes or anywhere. Certain species like termites, carpenter ants and wood-boring beetles may cause serious damage to one's property and cost one a lot of money.

2.2 PEST CONTROL GOALS.

2.2.1 Pest Prevention

Pest prevention refers to keeping a pest from becoming a problem. It is a goal if pest's existence or abundance can be predicted priory. Continuous pests are defined as very predictable pests. Periodical and potential pests are usually forecasted if one knows the conditions that will favor their existence as pests. Such an example is some plant diseases happen only under particular environmental conditions. If such conditions are existent, one can take measures to avert the plant disease organisms from harming the desirable plants (Fitzhugh, 2012).

Pest prevention is critical in the pest control strategy. If one waits until there is evidence of pests in the area, most likely the pests may already be there in large numbers. There are three rules pertaining to pest prevention that should be upheld: Deny pests access to the operation; deny pests food, water, and shelter; Work with a licensed Pest Control Operator (PCO) Instruction Notes

2.2.2 Pest Suppression

Pest Suppression has the purpose of reducing the number of pests to a level where the harm they are causing is acceptable. Thus, if a pest is detected, aims for undertaking suppression and prevention are made. Suppression of pest populations is often essential so as to avoid economic loss if prevention and avoidance methods are not successful. There are four primary suppressive methods that farmers or individuals employ: cultural, physical, biological and chemical control. Cultural practices may be such as narrow row spacing or optimized in-row plant populations, alternative tillage approaches such as no-till or strip-till systems, use of cover crops or adopting plant allopathy. Physical suppression methods can include cultivation, setting baits for some insects, and regulate temperature. Biological methods can be such as disruption of mating for insects. This is most suitable when it is necessary to control a stubborn pest (Fitzhugh, 2012).

2.2.3 Pest Eradication

Pest eradication refers to obtaining obviate each species from a region to that recolonization is unlikely to occur. Obliteration is an uncommon goal in outside pest circumstances since it's tough to attain. Usually, the goal is a hindrance and/or suppression of pests. In indoor areas, eradication may be an additional usual goal. Closed areas typically are smaller, less complicated, and higher simply controlled compared to outside areas. In most closed areas, like schools; workplace buildings; health care, food processing, and food preparation facilities, certain pests cannot or wouldn't be condoned under any circumstance. The eradication method involves 3 main activities including surveillance, containment, and treatment & or management measures. Surveillance refers to fully investigating the progress of pests; containment is preventing the pest infestation; Treatment involves elimination of the pest once it's found. Once eradication ends, absence of pests ought to be determined (Fitzhugh, 2012).

2.3 HYGIENIC METHODS OF PEST CONTROL

2.3.1 Clean All Surfaces after Eating

One should clear up any food debris if you use baits, so that the pest is attracted to specifically only the bait. If one had treated for any rodents such as mice or rats, it is necessary to clean up after eating. Occasionally, homes get invaded by mice and rats commonly due to something that attracts them. Most often, it is food and water. Leaving food in the open, or keeping it over the night or keeping in easily accessible containers will attract rodents. If onc owns a pct it is important not to leave food over the night. Cleanliness should be the topmost priority in order to keep pests away from any establishment. Kitchens and eating areas need to be always clean and don't allow dirt to accumulate anywhere in the home and even dustbins should be emptied daily.

2.3.2 Proper Disposal of All Food Scraps, Crumbs and Peels

Garbage containing food scraps should be kept in strong compact-covered garbage bins and the garbage should be emptied regularly from the area. When waste bins are not properly cleaned or closed, pests could get in and take dominance of the overflowing of food inside. Moreover, if the waste bin is not leak-proof, liquids may penetrate underneath and lead to breeding grounds for flies and other insects.

Methods for proper disposal may include: choose a one waste bin for all food scraps, leftovers or peelings: To prevent edible residues from piling up use of plastic liners and often cleaned: Rinse empty food containers thoroughly before throwing in the garbage: Storing food waste in a freezer until one is able to get rid of it. As long it remains frozen, it will not give bad odors that attract pests: Garbage should be in a can with a tight lid since rodents chew thick plastic containers (Fitzhugh, 2012).

2.3.3 Proper Dishes Washing

Dirty dishes among the sink could also be a goldmine for bugs. One ought to wash dishes once use and dispose of food scraps. Moreover, one should not let food sit and decay among the sink. A particular order ought to be followed once in cleaning dishes or utensils, that is; wash, rinse then sanitize. This is often terribly essential to keep every kind of pests removed from the dishes.

In case of use of pesticides then one determined to leave dishes or utensils within the room drawers, then it's necessary to clean every bit in heat water with dish soap before victimization for cookery. Cleaning up after meal preparation is crucial. One should have a positive mind that you simply don't have any food around the home. Uncovered food and food on the ground encourage the breeding of pests. One should simply make sure that the dishes are clean before retiring to bed.

2.3.4 Maintaining Proper Cleanliness

Crumbs and spills are the number one reason behind most pest invasion. To limit the areas pests are drawn to, one must always eat at selected areas for meals like dining or cafeteria. In these areas, trash ought to be kept in a very sealed bin and should be emptied frequently. Dirt and grime also can attract pests. One ought to confirm to wash and clean counters, tables, and any appliances like coffee makers or bread toasters. One ought to put food away and clean well after use since food remnants can attract flies and roaches. Trim shrubs and branches hanging close to windows or the perimeters of the building or home that encourage insects to invade residence. Confirm that windows, door frames, and alternative entry points are sealed and cleared of any debris. Moreover, repair sidings and decayed wood on or close to your building to stop pest invasions.

2.3.5 Store Food in Containers with a Tight Seal

Without correct food storage, one experiences inconvenient and long cuss issues, conjointly product harm and loss. Perishable food like meat and poultry ought to be wrapped firmly to take care of quality and stop meat juices from obtaining alternative food. To maintain quality, freeze meat in its original package, wrap the package once more with wrapping that's suggested for the deep freezer. Canned foods are safe indefinitely as long as they're not exposed to freezing temperatures. If the cans look okay, they're safe to use. Discard cans that are damaged, rusted, or swollen. Always use suitably sized containers for the contents. Avoid leaving unsealed food out, particularly fruits as they are magnets for flies. Moreover, check food containers for cracks, harm and loose seals to avoid leaks. Store all dried food products in containers with tight-fitting lids to stop product pests from going in and to assist contain any infestations that occur.

2.4 BIOLOGICAL METHODS OF PEST CONTROL

2.4.1 Use of Natural Predators

Biological management is a technique of dominant pests like insects, mites, weeds and plant diseases pathogens. Biological management assumes that natural predators are ready to suppress pests. Use of natural predators to manage pests refers to classical (importation), where a natural enemy of a pest is introduced within the hope of achieving control; inductive (augmentation).

Predators are chiefly independent species that directly consume an oversized variety of prey throughout their whole period of time. Proper identification of pests, distinctive pests from natural enemies, is crucial for effective biological management. Fastidiously observe the mites and insects on your plants to assist in making out their activity (Fitzhugh, 2012).

Bacteria, fungi and alternative micro-organisms will have these effects as a result of they compete for nutrients or space, they produce antibiotics or they merely eat alternative harmful micro-organisms.

Predators can be used preventively since they can make plants healthier and stronger. Once this happens, plants don't seem to be attacked by pests or diseases or are affected less by them. This sort of pest management isn't visible.

Some benefits of this technique include: the natural enemy will become established and this may manufacture long-run results. The chance of resistance is additionally a lot of lower since pests cannot build up resistance to being eaten. Natural pest management is extremely targeted and so an efficient way to manage specific pests. A disadvantage is that pests are never destroyed utterly as a result of the life cycle of a natural enemy and can thus never destroy the complete population.

2.4.2 Use of Microorganisms

They come from existent or genetically altered bacterium, fungi, algae, viruses or protozoans and may be used as alternatives to chemical pesticides. Microorganism e.g. fungus, virus or protozoan active ingredient targets various varieties of pests, though every separate active ingredient is comparatively specific for its target pest. For instance, there are fungi that control certain weeds and different fungi that kill specific insects (Fitzhugh, 2012).

These microorganisms give benefits such as: they need potency and safety for humans and different non target organisms; they leave less or no residue in food; they're ecologically safe, in order that different natural enemies are freed, resulting in the preservation of different natural enemies, and exaggerated variety in managed ecosystem. Microorganisms are extremely specific against target pests so that they facilitate the survival of helpful insects in treated crops. This might be the most reason that microorganisms are being developed as biological management agents throughout the last 3 decades.

The organisms utilized as microorganisms are primarily nontoxic and nonpathogenic to life, humans, and different organisms not closely associated with the pests (Fitzhugh, 2012).

The harmful action of microorganisms is usually specific to one cluster or species of insects, and this specificity implies that most microorganisms don't directly have an effect on helpful insects (including predators or parasites of pests) in treated areas. If necessary, most microorganisms may be utilized in conjunction with artificial chemical pesticides as a result of in most cases the microorganism is not deactivated or broken by any residues of typical pesticides.

2.5 PHYSICAL METHODS OF PEST CONTROL

2.5.1 Elimination of Breeding Grounds

Most disease vectors are insects, like mosquitoes, midges and flies. Mosquitoes need water for their immature stages, however, not all types of water are appropriate for all types of mosquitoes. Some need comparatively tiny pools of fresh, stagnant water (e.g. the arthropod genus protozoa infection vectors). Others like water in containers like jars, bottles, tanks, etc. Human activities, notably food production, eating, drinking, sleeping, excretion and wash, will promote the propagation of vectors and pests (Fitzhugh, 2012).

Defecation fields, for instance, should always be unbroken at a distance from preparation areas, due to flies and doable surface rain run-off. In most of tropical Africa, they ought to even be placed off from rain run-off to bathing or fishing waters, due to the chance of contamination with schistosomes. Another example is that the promotion of animal production and farming to scale back dependence on food distributions.

If hand pumps are put in poorly drained locations, the resultant waterlogging might end in mosquito breeding habitats, producing puddles containing water snails, or increase soil wetness sufficiently for hookworm transmission to become doable. Run-off water ought to thus be drained a long way away or allowed to percolate into the bottom in soak ways (Fitzhugh, 2012).

To prevent mosquitoes resting around homes, weeds and shrubs ought to be frequently cut. Rubbish ought to be removed/ burned a minimum of once every week to avoid the build-up of insect populations.

2.5.2 Poison Bait

Poisonous baits are chemicals that have deadly activity against targeted pests. The food sometimes acts as attractant. Food-based baits are an efficient and selective methodology of insect management. Typically, a bait consists of a base material referred to as a carrier (often grain or animal protein) and a toxic (most usual pesticides like organophosphates, carbamates or pyrethroids) and typically an additive (usually oil, sugar or water) to extend attractiveness. The toxic part of the bait can even be biological instead of chemical (Cohn, 2011).

Many baits aren't extremely engaging to the insect; however, instead operate to arrest. The insect isn't drawn to the bottom material, however finds it appetizing and can eat it once encountered. Therefore, baits are distributed by either ground or aerial broadcast application so as to make sure broad dispersion. Deadly baits are usually environmentally friendly, cheap strategies of insect management.

Poisoned bait could be a common methodology for dominant rats, mice, birds, slugs, snails, ants, cockroaches and alternative pests. The fundamental granules, or alternative formulation, contain a food attractant for the target species and an appropriate poison. For ants, a slow-acting poisonous substance is required so staff has time to hold the substance back to the colony and for flies, a quick-acting substance to stop additional egg-laying and nuisance. Baits for slugs and snails usually contain the molluscicide metaldehyde that's dangerous to youngsters and family pets (Cohn, 2011).

2.5.3 Field Burning

Farmers and ranchers manufacturing food and fiber for society are in an exceedingly constant battle against pests making a crop unsuitable for human use. A good example is nematodes living in soil. .

Burning is usually recommended as an impact procedure for a number of agricultural diseases and pests, burning isn't the only recommendation. In several cases, open burning could also be the foremost expedient control; in some cases, it should be the foremost economical management. Every case should be argued on the precise desires of the community. No blanket exception to a ban on open burning is required to regulate agricultural diseases or pests. Specific exceptions might, during a few cases, be excusable (Cohn, 2011).

Burning is additionally useful in removing weeds, preventing illness and dominant pests. For a few crops as well as rice, burning is that the most effective and effective thanks to management illness.

Burning is employed for persecutor and weeds management and lowers the necessity for supplemental weed killer and chemical treatments. As an example, the occasional burning of pruning and alternative vegetative detritus are vital controls for pests and illness within the plantation business. Burning is additionally utilized as a good means that of disease control on new pruning to regulate secondary pests within the older residue and for phytosanitary emergencies.

2.5.4 Trap Cropping

Trap cropping typically involves the early planting of border strips of a specific crop to draw in insects wherever they will be destroyed by pesticides. Trap cropping conjointly contains a tremendous potential to manage alternative crop pests. It will minimize the employment of pesticides and may be integrated with alternative IPM (Integrated Pest Management) tactics. The reduction in pesticides can end in less disruption to the atmosphere and to associated helpful organisms in cropping systems (Cohn, 2011).

There are various factors required to be taken into thought once planting a trap crop:

- Different insects like completely different entice crops: that entice crop you select depends on the pests you're making an attempt to entice. If the plant is not sufficiently engaging to the pest, then it will not be any use, therefore, it's very important to select the proper ones.
- The layout of the trap crop: for a few insects it's spare to plant the entice crop round the border of your growing space. Other area units are made more lasting to prevent and it should be necessary to interplant them to draw them off the most crops.

- Timing: Most insect invasions happen at a selected time of year. It's vital to own the entice crop already well established by the time pests arrive.
- Beneficial Insects: trap crops area unit only a part of smart organic pest management and wish to be balanced with adequate companion planting of flowers to draw in helpful insects like and ladybugs that kill the pests.

2.5.5 Traps

Insect traps are used to monitor or directly cut back populations of insects or alternative arthropods, by trapping individuals and killing them. They generally use food, visual lures, chemical attractants and pheromones as bait and are put in order that they are doing not injure natural enemies or humans or lead to residues in foods or feeds. Visual lures use light-weight, bright colors and shapes to draw in pests. Chemical attractants or pheromones could attract solely particular sex. Insect traps are typically utilized in pest management programs rather than pesticides, however, are a lot of typically wont to consider seasonal and spacing patterns of pest prevalence. This information could then be utilized in alternative pest management approaches (Cohn, 2011).

There are numerous benefits of insect traps as a way of pest control:

- Traps work twenty-four hours each day. Even if one scouts for insects every day, one will not see nocturnal insects that damage the plants
- Traps typically target adult stages of pests. These seem long before the caterpillars or grubs that kill plants. Therefore, traps facilitate early detection of pests before they cause injury to crops, and may assist you to correct bound issues
- Traps will indicate drawback areas, or hot spots, requiring corrective action or spot treatments while not having to treat all of your plants, and so facilitate cut back the employment of pesticides
- Trapping programs offer the knowledge required to base and document your management decision

2.6 CHEMICAL METHODS

2.6.1 Insecticides

Insecticides are pesticides that are developed to kill, harm, repel or mitigate one or additional species of insects. Pesticides add alternative ways. Some pesticides disrupt the nervous system, whereas others target exoskeletons, repel them, etc.

Humans have earned vital advantages from several uses of pesticides, including increased yields of crops thanks to protection from defoliation; hindrance of abundant spoilage of foods; and prevention of different devastating diseases(Cohn, 2011).

The transport sector makes use of herbicides and pesticides to check sports pitches, cricket grounds and golf courses. Pesticides defend buildings and picket structures from injury by termites and wood-boring insects.

However, when pesticides are used, they are doing not continuously keep within the location wherever they're applied. They mobile within the setting and sometimes move through water, air and soil. Chemical quality results in pesticides obtaining contact with different organisms and cause damage. Pesticides have conjointly been shown to disrupt the balance of the food chain. In several things, once a chemical is employed, it conjointly kills non-pest organisms to hence drastically alter the natural balance. Additionally, to inflicting damage to life, pesticides that travel from their original location are well-known to cause damage to humans. Human exposure to pesticides has caused poisonings, the event of cancer and also deaths (Cohn, 2011).

2.6.2 Rodenticides

Rodenticides are pesticides that kill rodents. Rodents encompass not solely rats and mice, but also squirrels, woodchucks, chipmunks, porcupines, nutria, and beavers. Though rodents play vital roles in nature, they'll generally need management. They'll harm crops, violate housing codes, transmit sickness, and in some cases cause ecological harm.

Rodenticides give a fast and straightforward resolution to rodent management regardless of however giant an infestation could be. There are several advantages related to the use of rodenticides:

- Easy to use and place - Most rodenticides are easy to use and need very little to no assembly. Some bait stations come back able to use with the bait already placed within creating it terribly straightforward to use.

- Single-dose kills - several rodenticides give a one dose kill for humane rodent disposal. This implies the infestations are in restraint rather more quickly as compared to different strategies
- Cost-effective - Single dose kills mean a lot of product is out there for mice and rats that haven't nevertheless discovered the bait station. With rodenticides, one is able to control and eliminate many pests with less bait.
- Child-resistant - Some rodenticides have features suitable for house with kids. They're kid resistant and are designed therefore tiny fingers cannot reach the poison.
- Attracts rodents to a selected location - Some rodenticides attract mice and rats to where you have got placed your bait station. The bait attracts them due to their gnawing characteristics such as oats- rodents notice the formula delicious and keep feeding till they consume a lethal dose.

2.6.3 Fumigation

This refers to a chemical vapor or gas that, once free, penetrates objects or confined area lethal in concentrations that are fatal to pest organisms. This definition excludes aerosols where area unit particles suspended into the air, usually named as smokes, fogs or mists.

Fumigants kill by interfering with the respiratory system of the target pest. Molecules of some fumigants (for instance, dioxide or inert gases) replace chemical element molecules within the air; therefore, the pest-control action involves suffocative (asphyxiation) thanks to lack of chemical element. The killing action of a chemical is influenced by its concentration within the atmosphere, the length of time it stays within the atmosphere, and therefore the temperature and humidity of the area at the time of fumigation (Cohn, 2011).

There are numerous benefits of using fumigation:

- Fumigants are toxic to all varieties of life. Therefore, it's attainable to regulate all life stages of a pest.
- Fumigation is usually the fastest method of targeting infestation, saving time and money.

- Fumigants will reach wherever sprays, dust, aerosols, etc. cannot reach.
- Reduced residue issues in treated areas.

Nevertheless, coating comes with its disadvantages:

- The management achieved through the coating is temporary – there's no residual action from fumigants. Where untreated populations of the pests stay, re-infestation of the treated spot will manifest itself quickly.
- Fumigants are usually toxic and infrequently extremely unsafe to the device, requiring special precautions throughout the application.
- Fumigants must be maintained within the gas form for an amount of time to be effective, often requiring extra supervision.

2.6.4 Heat Treatment

Heat treatment gets rid of all stages of the life cycle of pests by killing eggs, larvae and adult insects in one treatment. The warmth eliminates pests like bedbugs and cockroaches by denaturing targets inside their bodies and interfere the layers of wax on the surface of the insect, inflicting dehydration. This method kills all life cycle stages of insects that make it terribly effective in coping with infestations.

There are benefits to using heat treatment for pest control including:

- It is non-chemical: Heat treatments won't result in any long health facet effects of exposed individuals; chemicals are often cyanogenetic if not correctly vented before re-occupancy or if proper procedures don't seem to be followed.
- Minimal disruption: Single, fast and economical treatment but requires daily use. Treated areas are often used directly subsequently.
- Protects one's business reputation: Pest issues are solved speedily and totally, from early detection to speedy response.
- Minimizes costs: Mattresses and upholstered pieces of furniture are often treated, saving one the price of disposal and replacement.
- Reduces the employment of pesticides: Non-toxic and pesticide-free possibility, creating it particularly appropriate for sensitive environments like food preparation areas and hospitals.

- Real-time observation: Wireless sensors and computers are accustomed to remotely track temperatures in numerous areas throughout the area being treated. A close report is generated at the conclusion of the treatment.

2.7 BENEFITS OF NATURAL METHODS

2.7.1 Minimized Safety Issues

Biological management leave behind no long-lived residues. They do not leach into groundwater or produce resistant strains of insects. Biological management minimizes environmental, legal and public safety issues.

2.7.2 Increased Selectivity

Often a host-parasite or prey-predator relationship may be a specific one that does not have an effect on different organisms - a profit for the setting. As an example, it is vital to use the right species for the insect you wish to regulate. Biological-control nematodes are not parasitic on their hosts. They enter the insect through a natural body gap then eat a bacterium they carry with them that multiplies within the host. Nematodes have to be compelled to be the correct size to enter the actual insect and should have behavior that enables contact with the insect to start with (Cohn, 2011).

2.7.3 Environmental Safety

Another advantage of biological management is that pests are unable to develop resistance (Weeden and Shelton, 2005). For instance, we have a tendency to imagine that a good management of a pest by a natural enemy may cause sturdy choice on the pest thus may develop mechanisms of escape or tolerance to attacks by the management agent, breaking down biocontrol system but, as Van Emden (2004) mentions, "we apprehend of no cases wherever antecedently roaring biological management has unsuccessful attributable to choice for resistance".

2.8 DISADVANTAGES OF NATURAL METHODS

Biological management may be fickle. Ultimately, you cannot control no matter the natural predator you set loose in an ecosystem. Whereas it's speculated to manage one pest, there's continually the likelihood that your predator can switch to a unique target - they may decide to feed on your crops rather than the insects infesting them. Not only that, however in introducing a brand new species to an environment, there runs the danger of disrupting the natural food chain (Cohn, 2011).

It's a slow method. It takes loads of your time and patience for the biological agents to figure their magic on a blighter population, whereas alternative strategies like pesticides work offer immediate results. The side of the present is that the long-run result biological management provides. If one wants to eliminate pests, biological management isn't the correct selection. Predators will solely survive if there's one thing to eat, thus eliminating their meal population would risk their safety. Therefore, they'll solely scale back number of harmful pests.

While it's low-cost within the long haul, the method of really putting in place a biological system may be expensive like developing a thriving system like it's going to be valuable to research and high level of skills and initial found out costs.

2.9 ADVANTAGES OF CHEMICAL METHODS

The chemical methodology provides a quick way of eliminating pests. The chemicals are factory-made in a manner that allows them to kill pests in the shortest time potential. There are chemicals that stop pests' revivification. Using chemicals will facilitate a farmer to eliminate the whole pest population virtually outright by merely administering the chemical within the setting occupied by the pest. A major advantage of chemical management is its potency. Most chemicals act in no time and once hand-picked properly they're extremely effective in eliminating the pests (Cohn, 2011).

In some cases, chemicals are the sole ways that may be effective in dominant a particular pest. This is often typically the case wherever all alternative means that have failed or might not be sensible. Thus, chemicals will save crops from a lot of infestation as a result of their simply accessible within the native outlet or maybe on-line in contrast to some ways like biological ways.

Chemicals also are obtainable in several varieties. Dominant pests become a lot of economical once the chemical technique is employed. this is often as a result of a farmer wants simply to spot the persecutor then select a chemical that may kill that individual persecutor while not damaging useful insects. Additionally, chemicals are simple to use in dominant pests. Most chemicals accompany directions on a way to use them in controlling the target pests. This suggests that a farmer will continuously combine the correct amount of water with the chemical then spray the plant (Cohn, 2011).

2.10 DISADVANTAGES OF CHEMICAL METHODS

Chemicals are harmful as a way of dominant pests because they promote pests' evolution. Pests that are immune to the chemical develop resistance to that specific chemical and it will no longer be utilized in controlling them. If the developed resistance is genetic, the future generation won't be controlled using that chemical. Additionally, chemical management results in non-target effects and toxicity. Several chemicals kill even useful insects like bees and natural predators. If ill-used, chemicals may intoxicate and even kill humans.

Promote evolution: The use of chemical pesticides stimulates the pest to develop resistance to the chemicals employed in the pesticides. The resistance relies on alterations within the genetic science of the vermin and each future generation becomes progressively pesticide-proofed. This works alright among rodents that turn out many litters per annum.

Resurgence (non-precise targeting): Resurgence happens once the use of pesticides affects the environment and disrupts the organic pest management. The foremost tremendous aspect effects of all are once important insect species like bees get killed by casualty by chemical pesticides. Another is once pesticides are sprayed they kill natural enemies as well like parasitoid wasps and their pest hosts, no longer have a natural enemy and begin to outbreak and utterly overwhelm entire agriculture.

Persistent organic pollutants: POP, referred to as persistent organic pollutants, are extraordinarily dangerous to the non-target organism however additionally an effect on the health of individuals has by inflicting cancer, sterility or issues to the system.

CHAPTER 3

Classification of Pesticides

CONTENTS

3.1 INTRODUCTION

Pesticides are classified in different ways. This chapter was prepared to offer detailed information on these classifications as discussed below.

3.2 CLASSIFICATION BASED ON PESTICIDE FORMULATION

3.2.1 Liquid Pesticides

These pesticides usually have active ingredients that are not easily soluble in oil or water. The formulation is usually mixed with water or other agents that are applicable. It is easy to handle and does not clog spray nozzles and thus only need moderate agitation. However, farmers have evidenced acute poisoning from this form and it is associated with fatal accidental poisoning by pesticides if not handled with care. The choice of the formulation can also minimize hazards. Some formulations are less easily absorbed. In some formulations, the solvents used actually constitute the major hazard. Benzene, for example, is very toxic and should be avoided. Some formulations have been shown to be safe as pastes as compared to liquids. Liquid concentrates of parathion are safer when compared to the wettable powders of the same compound. This is attributed to the higher respiratory exposure experienced with the solid formulations.

Figure 3.1: Liquid pesticides.

Source: https://www.packrusa.com/industries/agrochemical/

Compared to other safer formulations are granules, baits, slow-release pesticides and micro-encapsulations. Wettable powders are more hazardous than liquid concentrates during the mixing. The efficacy of spraying different formulations has been found to be less than 0.1% of the applied dose of foliar sprays. A large proportion is suspended in air and could be a source of contamination. Studies on acetylcholinesterase levels in whole blood before and after spraying indicated that there is little difference in enzyme levels before and after spraying. The cloths that farmers wear act as protective clothing. They have been tested to confirm that they would not be easily penetrated by liquids (Soloneski, 2014).

3.2.2 Gel Pesticides

These designs are typically utilized for two primary applications, which is insecticide baits and for application on-skin insect repellents. Without gels, farmers would have to be exposed to their normal spraying operations. This is because most farmers hardly wear protective clothing. It has been concluded that protective clothing may prevent exposure which would be associated with depression in enzyme levels of the exposed farmers (Soloneski, 2014). Extraction involves the removal of pesticides from a large bulk of sample matrix and the method and solvents used depend on the type of substrate and final method of analysis. Non-bound residues are extracted directly by organic solvents while strongly bound residues are first released from substrate media by procedures often requiring derivatization, for example, silylation for phenolic metabolites of organophosphorus insecticides. A final step in extraction involves subjecting the sample extracts to a cleaning procedure to remove co-extractives that interfere in the analytical determination and instrument performance. Various clean- up procedures have been used which include column chromatography, acid and base treatment, gel permeation chromatography, silica gel and florisil clean-up. Challenges include use of leaking sprayers and unlabeled chemicals, application of pesticides without wearing protective clothing, improper storage of pesticides, and use of higher than recommended dose, use of chemicals whose use has been discontinued or banned both in different countries and in other parts of the world and improper disposal of pesticides and their containers. In other parts of the developing countries which have documented that pesticide handling and usage practices are poor hence exposing man and his animals to hazardous pesticides. Most farmers and agricultural extensionists normally use gel pesticides in their agricultural activities.

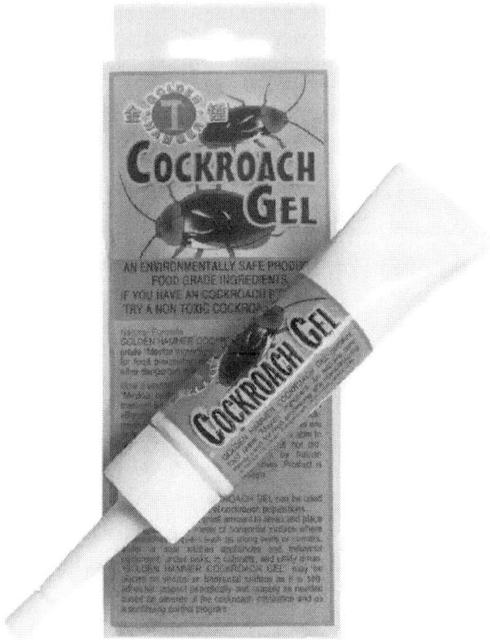

Figure 3.2: Gel pesticide.

Source: https://diypestking.com/product/cockroach-gel/

3.2.3 Paste Pesticides

A paste is a thick, soft and sticky material that is created through mixing a powder with a liquid to make a glue-like substance. It is volatile on rodents and insects since it is tamper-resistant. Paste pesticides, when used properly; enhance health and agricultural production with minimal side effects (Soloneski, 2014). They contribute directly to health through the control of vector-borne diseases like malaria, bilharzia, trypanosomiasis and leishmaniasis. It is estimated that over 10 million lives are saved per year in Africa through mosquito control for malaria alone. Pesticides also contribute directly to the economy via increased food production and through the protection of many materials during storage. Notwithstanding their importance, paste pesticides pose significant environmental and occupational health hazards throughout the world.

If the compound is used at any stage of food production, residues of it or its derivatives may persist in food, thus, exposing the wider population, although usually at much lower levels than chemical and agricultural worker exposure. However, the accidental contamination by paste pesticides of foods consumed by man has resulted in high morbidity and mortality rates in some cases. Domestic and wild animals may also be affected depending on how and where the compound is used and its persistence after use, but this is usually accidental. Pests may also develop resistance to paste pesticides. The hazard caused by a specific paste pesticide is often proportional to the amount used. Therefore, information on production and use constitutes an important factor determining both benefits and hazards. However, there is also wide variation in the toxicity of different compounds. The total side effects associated with paste pesticide use were estimated at 839 million dollars in the U.S. alone in 1980.

3.2.4 Chalk Pesticides

Chalk pesticide is used for getting rid of pests and it contains an active toxic agent. It is applied in cracks and water access routes. It destroys pests such as cockroaches and is very cheap and efficient. It is particularly toxic and is estimated to cause 40% of all pesticide-related illnesses reported to United States poison control centers (Soloneski, 2014). Chalk pesticides have been associated with serious adverse effects in birds, man and animals. In Europe and America, some researchers have reported that insecticide residues in birds caused harmful effects or even death. These include thin eggshells with low calcium content liable to breakage and reduced reproduction. Other suggested toxic effects in man and animals include carcinogenicity, teratogenicity, immunosuppression, embryotoxicity, infertility and birth defects. Spontaneous abortion and congenital malformations have also been reported in women exposed to organic pesticides. In other types of pesticides such as sprays which may be harmful, there have been cases of increased premature births and congenital malformations for pregnancies in floriculture workers who applied high levels of pesticides. Chalk pesticides also pollute the environment. In some countries, researchers reported that chalk pesticides were associated with the development of chronic pathology and temporary disability among female beet growers. In Sri Lanka and Malaysia, symptoms of chalk pesticide poisoning were reported in 7% of agricultural workers; chalk pesticide poisoning in Costa Rica between 1980 and 1986, with special attention to agricultural and chemical workers estimated that over 3000 persons were hospitalized due to chalk pesticide poisoning with 400 deaths. In East Africa, most cases remain unreported but a study by some workers on the status of chalk pesticide usage in this region showed that poisoning may be contributing a great deal to most illnesses in our hospitals.

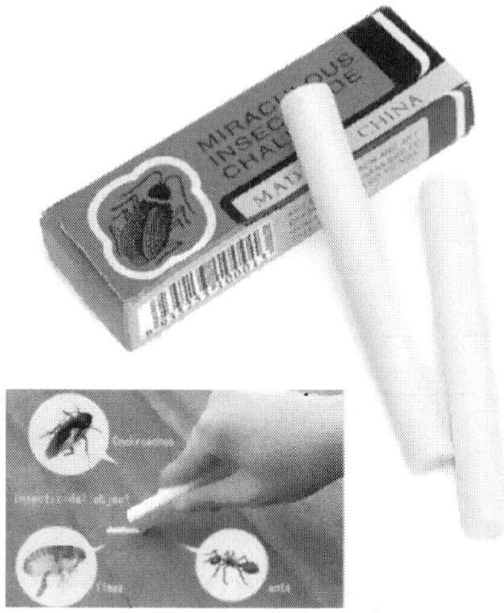

Figure 3.3: Chalk pesticide.

Source: http://lnnk.in/dZaw

3.2.5 Powder Pesticides

The powder formulation is a finely grounded mixture of materials that are intended for dry use. It should never be water-mixed. They are typically utilized for interior perimeter and wall treatments and also crop-dusting. An example is pyrethrin which is one of the oldest known, naturally occurring insecticides. Natural pyrethrins are extracted from *Chrysanthemum* spp. The flowers of these plants have been dried and powdered since the beginning of the 19th century for control of household pests. The first such products were imported over 100 years ago from Dalmatia and Iran. Natural pyrethrins have low mammalian toxicity and good knockdown effect. Different naturally occurring pyrethrin products have been classified as pyrethrin I and II, cinerin I and II and Jasmolin I and II based on electrophysiologic responses, clinical signs and chemical structure. Powder pesticide use within irrigation schemes is quite intensive with over 85% of farmers using these pesticides on either crops or animals. Previous studies carried out within the scheme identified several unsafe handling practices for powdered pesticides and it's advisable to have protective gear such as masks and gloves. The efficiency of different application methods showed that the proportion of applied dose taken up by biological target has a variation. When it contacts water it has been found to be least effective. To establish some safety guidelines, it is advisable to apply the powder without any other spraying activity to avoid weakening it.

3.2.6 Granules Pesticides

These are hard and dry particles that usually consist of materials that are porous and have active ingredients. The active ingredients are usually more than the dust in the formulation and are quite safe to apply. Insect infested fruits, vegetables and root crops fetch low prices in the market. In general, the use of pesticides is thought to reduce food prices. For example, a study commissioned in the United States to assess the importance of pesticides and fertilizers, estimated; if chemical use was stopped, there would be an increase in the annual food budget for the U.S family.

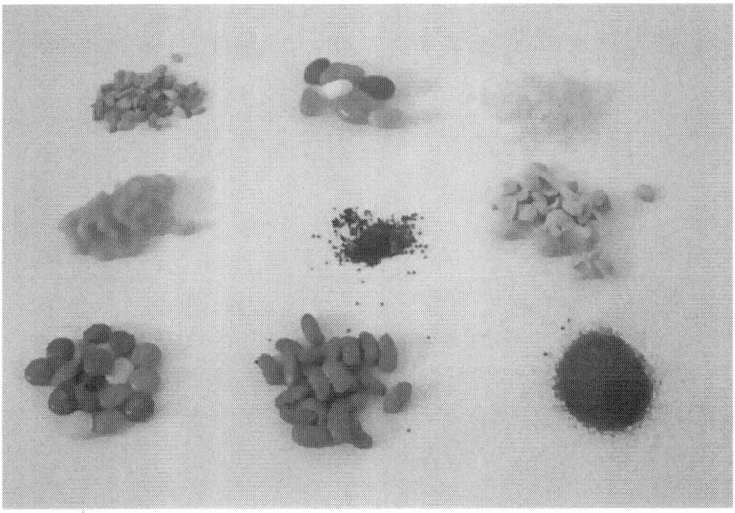

Figure 3.4: Granules pesticides play a major role in controlling pests and diseases and allow producers to secure markets for their produce.

Source: https://www.sciencebuddies.org/science-fair-projects/project-ideas/Phys_p091/physics/candy-waterfall-solids-flow

The use of orthodifolatan, for example, a fungicide previously used extensively in the control of coffee berry disease allowed different countries to be major coffee exporters in the 1970s and 80s. It was used to control the spruce budworm in North American forests. Also, in the last century, a fungal disease of American chestnut trees was controlled with fungicides. Likewise, in different countries, the cypress aphid was partially controlled by spraying with systemic insecticides. Granule pesticides provide protection against attack from bacterial, fungal and insect agents and has been practiced for centuries. Granule preservatives are needed to ensure long-term preservation purposes and short-term protection during storage. Another area in which granule pesticides have had a major impact is in public health. Vector-borne diseases of man have been a major problem for centuries with diseases such as malaria having a great impact on mankind throughout history. Granules pesticides have been used for protection against vector-borne diseases of man (Ohkawa, Miyagawa, & Lee, 2007).

3.2.7 Pellet Pesticides

Pesticide pellets are another formulation that has active ingredients that provide protection against pests. Current technology in application equipment may lead to efficient pest control with drastically reduced pesticide rates. The advantages of pellet pesticide application methods have been reported especially against rats. Pellets have various advantages over conventional pesticide techniques particularly:

- Reduced drift
- Controlled droplet size and velocity
- High recovery index (ability to land on the target plant)
- Low rates per hectare
- Maintenance free
- Low energy consumption

Field trials on the control of birds on cotton showed that the pellets were five times as effective as conventional pest control techniques for the same active ingredients used. In addition, the amount of diluent needed was ten times less than in conventional techniques. Various workers have suggested ways in which pesticide usage may be reduced while effective pest control is maintained. Knowledge of the behavior of the pest, crop and environment and their interactions is needed to better target pest control strategies that have both acceptable safety and cost.

For example, pests can best be controlled if carefully timed suppressive applications are targeted at weak points in the pest's life cycle. To achieve this, life tables of both the pest and the crop have to be prepared (Matthews,

2008). Pest control is then directed at that stage of the crop when it is most susceptible. For example, the cabbage crop in Ontario, Canada, is affected by numerous insects but the most serious one is the cutworm. Studies on the life cycle of cabbage have shown that the critical periods in control are in the early stages when cutworms cause the most mortality of plants and in the heading stage due to damage by caterpillars.

3.2.8 Baits Pesticides

Unlike classification by chemicals designed to kill, reduce or repel pests, baits are substances that lure pesticides using food-like mixtures that attract them and are edible consequently poisoning them. They are very crucial for rodent control including rats, mice, flies, ants, cockroaches and various insects. An effective pest control strategy would, therefore, be directed early in the life cycle of the plant and later during heading. The timing of the spray period needs to consider other factors than the pest. For example, spraying should be timed to minimize the mortality of beneficial insects. Thus, the life cycles of parasites, predators and beneficial insects should be studied and the pesticide can then be applied stage when harmful effects to pests are maximized and to beneficial insects minimized; this application of non-persistent pesticides, strategy status and outlook of pesticide application. Pest control programs fail in achieving their objectives due to factors related to application equipment and its calibration, usage and maintenance. These shortcomings result in high costs, waste of pesticides, intoxication hazards, destruction of beneficial organisms, crop damage, excessive pesticide residues and environmental contamination.

Figure 3.5: The correct choice of bait plays a vital role in efficient application.

Source: https://www.aliexpress.com/item/33011025158.html

In many cases, farmers cannot identify the specific pest they are controlling. Substituting for the conventional routine treatment schedules would reduce drastically the amount of pesticides applied. It is important to note that 100% control of most pests is not required to prevent economic loss and the objective should be to prevent pest damage from exceeding economic benefits (Ohkawa, Miyagawa, & Lee, 2007).

3.3 CLASSIFICATION BASED ON THE TYPE OF THE PESTICIDE

3.3.1 Insecticides

Insecticides are used in the control of insects. Despite their benefits, the use of insecticides has been associated with a number of problems. After the discovery of the synthetic organic pesticides and especially their manufacture in the 1940s, there was a shift in insect control techniques with an emphasis on the use of chemicals. Sanitation, mechanical and biological control methods were essentially abandoned in favor of chemical pesticides (Soloneski, 2014). Due to the heavy and repeated usage of insecticides, secondary pest problems developed. Insects such as bollworms in cotton and rollers on apples and peaches increased as their insect predators such as boll weevils on cotton, and codling moths on apples were eliminated by insecticides. Thus, in some cases, all insecticides did was to shift the major and minor pest species.

The selective application of pesticides to coincide with behavioral characteristics of the pests has also been used effectively to reduce pesticide usage. Insecticides that are toxic to honeybees have been recommended for application after bloom is completed or in the evening when the bees are not foraging on crops. Similarly, insecticidal paints have been used to control mosquitoes. The adult mosquito usually rests on walls and ceilings after ingesting blood from humans. A small percentage of active ingredient of lindane is incorporated in paints prior to their application on walls.

The degree of efficiency in pesticide application is widely recognized as a factor that may influence the amount of pesticide used in pest control. In crop spraying operations, for example, the pesticide is usually aimed at a false objective since many operators identify a target area, not pest, and try to provide an even cover of pesticide over the total plant surfaces. Researchers argue that the pest occupies a special niche in the crop environment and this should be the target area. The nature of the biological target would then determine the droplet size and therefore also the pesticide formulation required. Significant amounts of insecticides that animals are exposed to for tick control purposes are absorbed through the skin after dermal application on cattle.

In practicing agriculture, crop yields have often been diminished due to insects and rodent damage, plant disease and weed competition. In addition, man has been a host for a variety of disease-causing organisms vectored by pests. Thus, through the ages, there has been great interest in compounds and methods to control or eliminate pests and diseases. As early as 1500B.C, insecticides preparations against lice, fleas and wasps were used. Around the eighth century B.C, Homer composed a poem about the "rinsing" of a cup with sulfur. Systematic studies on the use of chemicals to control pests, however, date back to the mid-19th century. One of the most notable of these was insecticides such as Prevosts which demonstrated the fungistatic effect of copper sulfate. Additionally, uses of mixtures of copper sulfate and lime have been discovered to be effective in controlling downy mildew of grapes (Soloneski, 2014).

3.3.2 Herbicides

Herbicides are used for plants. They have many benefits in the control of pests but also pose different risks. For example, pyrimidilium herbicides are delayed skin irritants and methyl bromide produces gross blisters on skin. There are various types of allergic dermatitis due to herbicides. Allergic dermatitis has been observed on exposure to phenoxy herbicides and anilides. Contact allergic dermatitis has been observed with a number of pesticides: benomyl, lindane, malathion, simazine, trifonate and zineb. The third type of allergic dermatitis is provoked by sunlight and referred to as photo-allergic dermatitis. Compounds that have been implicated in this are atrazine, benomyl, thiram and zineb. In what appears to be a case of causal association, the length of arsenic exposure was related to lung cancer incidence and later reported deaths of workers exposed to herbicides. Trichloro phenoxy acetic in the United States; based subsequent case -control studies confirmed increased risk to cancer on exposure to chlorophenoxy acids. It has been suggested that this increased risk could have been caused by contamination of these herbicides with chlorinated dibenzodioxins. In a case- control study in Southern Sweden, a fivefold increased risk of soft tissue sarcomas in people exposed to chlorophenoxy and chlorophenol herbicides was estimated. Other effects of pesticides in humans have also been observed. Pyriminil, a substituted urea molecule also used as a rodenticide was found to cause severe diabetes and chronic neurological disorder. Another rodenticide, warfarin, is an anticoagulant and increases hemorrhage during gestation. When administered as a drug, it is teratogenic in humans although there is controversy about this.

Among herbicides, paraquat is known to be acutely toxic on ingestion. In addition, a case occurred in which paraquat was found to be toxic on skin contact as well. Skin contamination from a leaking sprayer caused the death of one worker and reduced lung function in several others. Extraction involves the removal of pesticides from a large bulk of sample matrix and the method and solvents used depend on the chemical and physical properties of the pesticide, type of substrate and final method of analysis. Non-bound residues are extracted directly by organic solvents while strongly bound residues are first released from substrate media by procedures often requiring derivatization for example, silylation for phenolic metabolites of organophosphorus herbicides (Ohkawa, Miyagawa, & Lee, 2007).

Herbicides are an expensive input in crop production. A major reason is the high cost of development of new compounds and the need for rigorous testing to satisfy regulators of their safety for all users. Recently, the cost of developing a pesticide has been estimated at $20-30 million as compared to $1.2million in earlier years. These costs are for pesticides that reach registration stage. Smaller but substantial amounts are spent on large number of compounds that do not reach this stage. These costs are incorporated into the wholesale and retail cost of marketed pesticides. This was not expected because the tack is on the opposite side of the spray drift and not in contact with sprayed plants that may have pesticides on them. Exposure may occur, however, from a leaking pump, cap or from pavers without a cap.

3.3.3 Fungicides

Fungi have been shown to reduce yields as well as the quality. Due to this reduction of yield by fungi, various fungi control methods are used. The most common are cultural (using forked hoes, slashing, mulching, intercropping and high-density plantings) chemical (fungicides) and integrated method that is combining both cultural and chemical fungi control. Although application of fungicides in coffee has substantial yield advantage, it is not ideal where soil erosion is likely to occur. Slashing and use of fungicides are a better alternative under such conditions.

Every effort must be made to ensure that only healthy planting material is used as this is the most effective way of preventing disease and attack by pests. Several different types of infection start in the shoot. Therefore, the planting material should eventually be dipped in fungicidal solutions to protect it against mealybug and fungal diseases. The choice of chemicals used is specific to different diseases or plantations.

Studies in which specific effects are known or suspected are generally more informative. Examples of specific disease effects have been demonstrated. For instance, male fertility-as measured by low sperm counts and increased production of reproductive hormones is associated with long-term exposure to dibromo-3- chloropropane (DBCP) (Soloneski, 2014).

There are many reported cases of plant -pathogen resistance to fungicides. Fungicides to which important resistance has developed include benzimidazoles, dicarboximides, pyrimidines and organotins. The accumulation of pesticides in the global ecosystem and their effect on species of birds and fish caused a major problem in many parts of North America in the 50s and 60s and was brought to public attention. The fungi species are classified into annuals and broad-leaved.

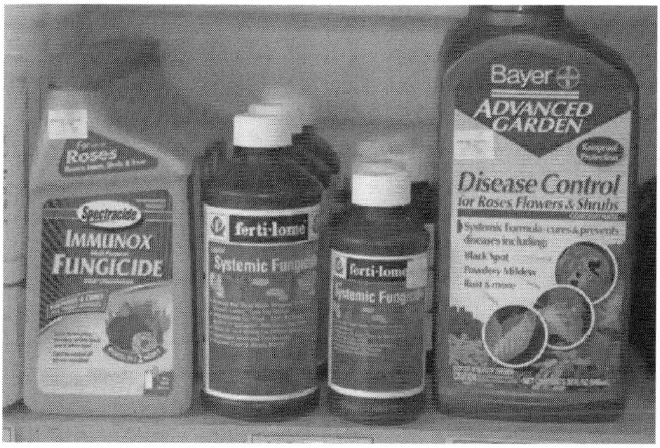

Figure 3.6: Fungicides which are classified into foliar and soil, contact and systemic have been used extensively in farming.

Source: https://pnwhandbooks.org/plantdisease/pesticide-articles/fungicides-disease-management-home-landscape

The use of one type of fungicides leads to the development of tolerance of some fungi. For instance, continuous use of gramoxone has led to tolerance in some fungi. The use of fungicides is comparatively cheaper than hand fungi, but profitability is influenced by changes in coffee prices. The use of soil, acting fungicide, followed by spot application of a contact fungicide is more economical in high rainfall areas than use contact fungicides alone. The use of low rates and volumes of recommended fungicides can be used to effectively control annual fungi in coffee. This has become possible through the development of low volume nozzles. For better control of annual fungi, with low rates, it has been recommended to control them at the early stages. The low rates and volume used would be a saving to the farmer and an improvement in environmental safety (Ohkawa, Miyagawa, & Lee, 2007).

3.3.4 Rodenticides

Increased agricultural production has been attributed to the usage of these chemicals in many parts of the world. However, undesirable environmental and health effects have resulted from rodenticides particularly when mishandled. Both ordinary people and agricultural workers may be exposed. Globally, the few studies carried out reported that rodenticides are frequently mishandled and storage practices are poor. The extent of household exposure has rarely been assessed. Elsewhere, household rodenticides exposures have been reported to occur. The most common source of exposure has been eating contaminated food. Reports abound of the fatal effects of eating food contaminated with pesticides.

Rodenticides poisoning is common due to the lack of skill in handling the products. In a past incident, it was reported that the use of dinitro-ortho cresol (DNOC) resulted in casualties requiring hospital treatment when the pesticide was used in dusting against rodents. The usage and handling of rodenticides are controlled by the Voluntary Precautions Scheme. The obligation to handle rodenticides safely is placed on the use and is voluntary. There are some acts put in place for safety, for example, the Control of Pesticides Act was enacted and was valid for all countries within the East African Community. After the break of the Community in 1977, a Pest Control Products Act was drafted and finally enacted in 1982. To administer the act, the Pest Control Products Board was launched in 1983 and started its operations in 1984. This Board is charged with the responsibility of regulating all aspects of pesticide usage in Different countries today.

Statistics are lacking in the usage of rodenticides before the formation of the Pest Control Products Board in 1984. Records available since then indicate that importations increased gradually from 7000 tons imported in 1984 (PCPB, 1986) to reach a peak in 1987 at 10,371 tonnes. From then on there has been a decline in tonnage to the lowest level of 3533 tons in 1993. The total figure for 1994 is unavailable but there has been a slight upward trend. Most countries import different rodenticides and other pesticides such as organophosphates, carbamates, synthetic pyrethroids and organochlorines as well as various classes' rodenticides. Plants such as potatoes usually vegetatively propagated by means of tubers are normally referred to as the 'seed'. The plant is herbaceous with spiral phylotaxis. The potato has a fibrous root system formed either by the seedling taproot or by adventitious roots in plants raised from tubers. The plant develops stolons from the most basal nodes below soil level and rodents such as moles may attack them (Subramanyam & Hagstrum, 2012).

The contamination of indoor air has often had serious consequences for occupants. In some countries, contamination has resulted from a planned application of rodenticides to control indoor pests. For the purpose of reducing rodenticides hazards, labels, containers and packages should have instructions on safe usage and precautions, symptoms of poisoning, first aid measures and advice to physicians in case of poisoning. Other aspects that are likely to contribute to reduced exposure are regulations on protective clothing, registration of less toxic pesticides, control of contamination during transport and the periodic medical examination of workers.

3.3.5 Fumigants

Legislation on fumigants production, use and storage has increased over time. One of the earliest legislations was a State Law of New York enacted in 1898 to regulate the sale of Paris Green which was the most important pesticide at the time. A number of regulatory schemes have been described by the voluntary scheme is based on mutual trust and understanding. It requires that those concerned should be responsible without any legal actions being taken if there is non-compliance. The preventive scheme of bids the production, sale or use of pesticides except under the laws stipulated. Such a system requires elaborate bureaucratic procedures and considerable resources. However, when enforced, hereby the government handles all pesticides and regional schemes, which may cover many countries in a region. After the enactment of a specific regulatory procedure, accompanying countries' regulations are needed to enforce the law. Such regulations need to cover issues such as registration, licensing, trade, labeling, packaging and worker safety. Careful wording of these regulations is needed for the day-to-day enforcement of the regulations.

A national organization is needed to implement fumigants' laws and regulations. A suitable person is usually the Minister of Agriculture. Resources would then determine the size of the organization. A scheme may start with one person taking care of quality control and with time other aspects of efficacy, residue control, evaluation of user and bystander hazards as well as toxicological hazards can be added. Reducing exposure certain instruments of the law may be used. Registration is considered to be the best tool for this. Registration can be a special tool in controlling fumigants. For a product to be registered for use, information on physical and chemical properties has to be supplied by the manufacturer. Information needed is biological efficacy, toxicological, environmental fate and effects, and due data.

Fumigants that are highly toxic, consistent or bioaccumulative are either not registered or their registration is canceled or suspended. For example, a large number of chlorinated hydrocarbon insecticides are either banned or severely restricted in both developed and developing countries. Licensing may also be used in the control of pesticides. Premises used in manufacture, storage and distribution of pesticides can be licensed. Licensing regulations stipulate the conditions which must be met before licensing. The emphasis is usually on worker safety and protection. The use of precautionary measures in pesticide premises can make it possible to manufacture even highly toxic pesticides without endangering the health of the workers.

In addition to the licensing of workers, pesticide applicators can be licensed to operate. The licensing is subject to certain training. In Canada, for example, all applicators are required by provincial law to be licensed. Other regulations to reduce exposure can control the destruction and disposal of wastes, storage requirements, protective clothing, monitoring of workers, fumigation of premises and dressing of seeds. In early studies, the number of person-years of exposure was counted and relative exposure levels established. Quantification of chemical levels in the environment was an added source of information. Such measures can be usefully supplemented by measurement of chemicals and /or metabolites in body fluids (Ohkawa, Miyagawa, & Lee, 2007).

3.3.6 Wood Preservatives

Wood preservatives may be used for fencing poles or other building materials. Contaminated wood either used in feeding troughs or fences is considered an important source of pesticide residues. For example, a study in France showed that some goats died suddenly after grazing on a bait of toxaphene sprayed on wooden structures which were been put up to deter owners from allowing goats to trespass into farms of predominantly cassava and sweet potatoes which are an important source of green forage for livestock. A dog that consumed blood from some of the carcasses also died. In another study organophosphate and organochlorine pesticides were associated with deaths of 125 out of 128 dairy cows in the coffee belt of Northern Tanzania.

The majority of the poisoning cases occurred after the animals had consumed feed or water contaminated with pesticides from wooden preservatives. This was consistently confirmed by demonstrating the presence of pesticides in both the host and ingested material. In another incident, cases of poisoning of livestock after grazing on pasture treated with organochlorine pesticide, heptachlor, were reported in cattle body fat after grazing land uniformly treated with up to 1.1 kg/ha of commercial heptachlor. Fuelled by fears of the harmful effects of chemical exposures, many epidemiological studies have and are being conducted to assess the health effects of such exposures (Ohkawa, Miyagawa, & Lee, 2007).

Figure 3.7: Wood preservative.

Source: https://www.shreejiwoodcraft.com/manufacture.php?val=4&id=

An important task in such studies is to quantitatively establish exposure. This is often difficult because of the relatively long exposure periods required to induce disease. Biological monitoring as an important method for obtaining data on chemical exposures was accepted at a symposium in Finland in the early 80s. In biological monitoring, physical, chemical and immunological tests can be used to test for the presence of life-threatening residues from the wood preservatives. In addition to defining exposure, disease outcomes have to be carefully definedas both acute and chronic effects can result. Wood preservatives exposure is the frequent non-specificity of symptoms. For example, an outbreak of febrile illness in newborns was observed and treated as an infection but was later found to be poisoning from Pentachlorophenol, a wood protectant.

These cases are likely to be only a small proportion of all known cases. Furthermore, many cases of morbidity go unreported. Information is lacking on actual pesticide handling practices that may explain how exposure occurs. Globally, it is evident that farmers regularly use pesticides and frequently mishandle them. Actual exposures have, however, not been quantified. It has been attributed to many cases of pesticide poisoning to occupational exposure in agriculture. When combined with data on usage and handling practices, such information would assist in identifying how exposure occurs in the field and around the home. This would help extension workers to advise farmers on better pesticide handling procedures to reduce the risk of the adverse effects of pesticides. Improved knowledge on the behavior of pesticides in the environment and in the target pest location, evaporation, movement in soil, plant uptake and general metabolism) can be used modeling and predicting environmental impacts in a way that has not been possible in causal analysis using simplistic information such as customary criteria. Pesticides with fewer non-target effects would then be preferred (Philip & Landrigan, 2018).

3.3.7 Insect Repellents

The possibility that safer pesticides may be developed as new technologies develop. An understanding of the mechanism of action of pesticides and especially the specific biochemical site involved in pesticide action is likely to lead to safer usage which is evident with the use of insect repellents. For example, die organochlorines, organophosphates and carbamates target the nervous system of the insect while species such as diflubenzuron, a substituted urea insecticide, target the insect cuticle and acts interfering with the synthesis of chitin, a much more insect-specific attribute. For example, general repellents or systems to avert insect feeding on crops are potentially useful and safer than current chemical pesticides. In addition, pheremones, growth hormones, plant-derived compounds, microbial systems and sterilization techniques are alternatives to conventional pesticides.

Genetically engineered plants and animals that are incorporated with genes for pest control are expected to be produced in future and will play a wider role in crop protection (Philip & Landrigan, 2018).

Integrated pest management (IPM) involves the use of multiple modalities in crop protection. The use of this approach gives chemicals a lower priority by requiring that they are used in smaller concentrations in combinations with other methods. Thus, improved agronomic practice, improved land use management, plant selection and ecological manipulations are aspects of integrated pest management that are likely to reduce the amounts and thus hazards associated with pesticide usage. It has been suggested that some of the problems associated with the use of pesticides are a result of unnecessary use. Traditionally, use instructions for protection are always provided to guide the user on safe practice. Insect repellants should be sprayed before the onset of disease. In the 1960s and 1970s, time application for insecticides in the United States before planting corn was a common practice. In many cases, farmers could not identify the specific pest they were controlling; substituting for the conventional routine treatment schedules would reduce drastically the amount of pesticides applied. There is 100% control of most pests is not required to prevent economic loss and that the objective should be to prevent pest damage from exceeding economic thresholds. Various workers have suggested ways in which pesticide usage may be reduced while effective pest control is maintained.

Knowledge of the behavior of the pest, crop and environment and their interactions is needed to better target pest control strategies that have both acceptable safety and cost. The most common reasons for use of pesticides in and around buildings are termite control, pest control in food stores and in vector control programs. Insect repellents applied in this way have been shown to persist in the air for periods of up to one year. For example, in houses treated with chlordane, there was little difference in the amount of pesticide. There are chemically induced problems and techniques for investigating them. The information would be available to all interested individuals as promptly as possible for immediate use in dealing with daily problems. Regulatory authorities may also publish leaflets on pesticide recommendations, handling precautions and other information relevant to exposure reduction (Philip & Landrigan, 2018).

3.4 CLASSIFICATION BY USE AND CHEMICAL STRUCTURE OF THE PESTICIDE

3.4.1 Pyrethroids

This is popular insecticide and it is an organic-based compound that is comparable to the natural pyrethrins similar to what is produced in pyrethrum. Initially, organochlorine pesticides, principally pyrethroids were used extensively to control malaria and more than 20 other human diseases. Because of the persistence of DDT and other organochlorine pesticides, they have been largely replaced by organophosphorous compounds, synthetic pyrethroids and carbamates in vector control. Pyrethroids are very common due to their effectiveness; the current trend in pesticide usage shows a decrease in the use of organochlorines with increased use of organophosphates and synthetic pyrethroids. Thus it is expected that there will be a gradual decline in organochlorine residues. Synthetic pyrethroids (cypermethrin) are being adopted at a global scale and their use has been efficient. The measurement of pesticides, or their metabolites various biological samples such as urine, feces, perspiration, blood, milk, hair and nails shows that it can be harmful and thus it is necessary to handle with care.

3.4.2 Organophosphorus

This is another important insecticide. It is an organic compound and it contains elements of phosphorus. The most common use of organophosphorus is the measurement of blood cholinesterase inhibition following exposure to its compounds. Organophosphorus compounds are the largest and most diverse group of insecticides and are popular in many countries globally. First reported in 1820 that phosphoric acid could react with alcohols but it was not until 1930 that Schrader discovered the insecticidal applications of such a reaction. Organophosphates are easily degraded hydrolytically, enzymatically and biologically and require low application quantities to obtain the desired insecticidal activity, thus, reducing the danger of undesirable residues in harvested products. According to the WHO Classification of Pesticides by Hazard, a large number of organophosphates are classified as extremely, highly or moderately hazardous. The most notable case of reentry poisoning was an episode involving one hundred workers exposed to parathion residues in peach orchards; therefore, you have to handle organophosphorus with great care (WHO, 2010).

3.4.3 Carbamates

This is classified as an insecticide. It uses the carbamic acid to kill insects the same as the way an organophosphate insecticide. The short-residual carbamates are high to moderately toxic and are therefore generally hazardous in small amounts. Past scholars' evidence that poisoning is common for compounds such as sulfur pesticides, carbamates, chlorinated hydrocarbons and organophosphates. Among these, the carbamates are the most commonly implicated in reentry poisoning due to their inherent toxicity and the relative ease with which the oxon metabolites are formed under field conditions. They are also absorbed through the skin. The level of exposure to an inhibiting agent can cause an increase in development of problems, therefore, make sure you have low- level exposure to carbamates and precautionary measures. Pesticides that have carbamates, when used properly, enhance health and agricultural production with minimal side effects; they contribute directly to health through the control of vector-borne diseases like malaria, bilharzia, trypanosomiasis and leishmaniasis among others.

3.4.4 Organochlorine

Organochlorine is a popular insecticide. It is used in pesticides and other substances; it contains chlorinated molecules. The regular use of organochlorines is credited to its persistence since they are most commonly detected based on the current trend in pesticide usage shows a decrease in use of organochlorines and increased use of organophosphates and synthetic pyrethroids. This could have an impact on reduced popularity of organochlorine on a global scale. It is expected that there will be a gradual decline in organochlorine residues. Since an organophosphate (dimethoate) and a synthetic pyrethroid (cypermethrin) are the compounds of interest. It is estimated that over 10 million lives are saved per year in Africa through mosquito control for malaria alone. Organochlorines have contributed directly to the economy via increased food and fiber production and through the protection of many materials during storage. Notwithstanding their importance, pesticides pose significant environmental and occupational health hazards throughout the world. In addition, 25 million agricultural workers in developing countries are estimated to suffer some symptoms associated with their use of or exposure (WHO, 2010).

3.4.5 Manganese compounds

They are compounds of manganese and contain different oxides. The use of any biologically active compound poses potential problems of toxicity. This can occur through direct contact with the manganese compounds during manufacture, formulation or use. If the compound is used in any stage of food production, residues of it or its derivatives may persist in food, thus, exposing the wider population, although usually at much lower levels than chemical and agricultural worker exposure. However, the accidental contamination by manganese compounds pesticides in foods consumed by man has resulted in high morbidity and mortality rates in some cases. The hazard caused by a specific pesticide is often proportional to the amount used. Therefore, information on production and use constitutes an important factor determining both benefits and hazards. However, there is also wide variation in the toxicity of different compounds. They are particularly toxic and are estimated to cause 40% of all pesticide-related illnesses reported to United States poison control centers (Hallenbeck & Cunningham-Burns, 2012).

3.4.6 Thiocarbamates

This is an important fungicide. Fungicides have been associated with serious adverse effects in birds, man and animals. In Europe and America, thiocarbamates insecticide residues in birds caused harmful effects or even death. These include thin eggshells with low calcium content liable to breakage and reduced reproduction. Pest and diseases were considered a major constraint with 85% of the fanners using chemical thiocarbamates on either crops or animals. Pesticide formulations were used or stored and included; another group of pesticides associated with skin irritation is the thiocarbamates. These products have been used for a long time and are generally regarded as harmless. Floriculture workers who applied high levels of thiocarbamates pesticides are cautioned to take precautionary measures. Most cases remain unreported but a study by some workers on the status of pesticide usage in this region showed that pesticide poisoning may be contributing a great deal to most illnesses. Therefore, it is important to handle them with great care.

3.4.7 Dithiocarbamates

Dithiocarbamates is another functional group categorized in organic chemistry. Dithiocarbamates are most commonly used pesticides. Different countries use them in control of fungi. Constant use has led to increased pesticide residues. Tomatoes from markets in different countries found dithiocarbamate residues to be above-set maximum limits. Flora and fauna of lakes have in the past have shown the presence of dithiocarbamates pesticides in different samples. Consequently, white pelicans and lesser flamingoes from lakes were examined and found to contain high levels of dithiocarbamates residues in their livers e.g. Lake Victoria had shown high levels of dieldrin. This corresponds to the widespread detection of residues in different fish species. Therefore, the dithiocarbamates residues have been poisoning along the food chain command. There is a need for assessing the knowledge and awareness of health care providers in the recognition and potential of pesticide poisoning. There is increased use of the dithiocarbamates credited to its efficiency and effectiveness s in pesticide control.

3.4.8 Cupric Salts

They are copper-based salts and past studies have revealed that cupric salt chemical residues in animals may result from substances deliberately applied directly to the animal, feeding on contaminated feeds (either intentionally or unintentionally) or from contamination of the environment in which animals are raised. Contaminated feed either as pasture or harvested crop is considered an important source of intake for pesticide residues. Common mistakes include improper dosage, use of improper compounds or formulations, use of treated seed as feed, improper spraying and improper storage and disposal of pesticide containers. The disposition of pesticides or their metabolites in animals and their products as a result of ingestion of treated feeds is not well documented. The major sources of pesticides that animals in different countries are exposed to be cupric salts used to dip or spray livestock, the feeding of sprayed forage, grains and other feeds. The main cupric salts are salt-like although development of resistance by pests has led to introduction of combinations with pesticides.

3.4.9 Tiabendazoles

It is an antiparasitic and antifungal agent. The harm caused by tiabendazoles pesticides is mostly due to exposure to animals which in most cases is accidental, as a result of feeding near areas. Tiabendazoles chemicals may reach animals by accidental consumption or through feeding of sprayed foliage. Persistent chemicals like tiabendazoles and industrial wastes tend to accumulate in the food chains, eventually exposing man and animals that are at the top of most food chains that have been recently sprayed. A study by some researchers in an intensively farmed area of African farms showed that a wide variety of fungicides are used on food crops including tiabendazoles in place of organophosphates, carbamates, phenoxyaliphatic acids, pyrethroids and organochlorines in some cases. Many compounds are stored in tissues or eggs and present in milk after administration to animals. However, it is unusual for tiabendazoles pesticide residues other than chlorinated hydrocarbons to be detectable in more than trace amounts in marketed food of animal origin.

3.4.10 Triazoles

Triazoles belong in the class of organic heterocyclic compounds composed of two carbon and three nitrogen atoms. Triazoles are harmful if care is not taken. Many third world countries do not have the expertise or resources to ensure that dangerous chemicals are safely used. For example, majority of the farmers apply triazoles without wearing any form of protective clothing and they rarely changed the clothes used during pesticide application which increases the chances of pesticide absorption through the skin. These practices result in over three million people being poisoned annually by pesticides. There is much evidence to show that there are certain stages of crop growth when the presence of weeds no longer affect the growth and yield of a crop adversely. Each crop has its own critical weed competition period. This is the period during the growth of the crop when the presence and competition of weeds are harmful to the crop.

3.4.11 Dicarboximides

It is a fungicide in agriculture and it includes iprodione. Dicarboximides are very effective and are available for crop protection and are used prior to herbicides. In fact, in some experiments carried out in India, dicarboximides use for weeding was found to be more economical where maximum net return extends from the time the shoot emerges above the ground to the time when maximum haulm growth is achieved after which foliage senesces. During this phase haulms and roots develop simultaneously. As the plant becomes autotrophic, its sensitivity to environmental factors, such as temperature, radiation, water and mineral nutrients increases. It was also found that although farmers were concerned about potential adverse health effects of pesticides they took little responsibility for handling them carefully. Farmers associated various complaints with handling and spraying dicarboximides, including chest pains, coughing and sneezing, colds, swellings, itchiness, dizziness, fatigue, diarrhea and uneasiness.

3.4.12 Dinitrophenoles

Dinitrophenoles is an oxidative inhibitor and antiseptic characterized by solid yellow crystals. Pesticides have been widely used in controlling pests and diseases in agriculture, forestry, veterinary and public health. They also play a major role in maintaining weed-free lawns, sports grounds and in clearing roadsides. Pest- induced losses due to plant diseases, arthropod and vertebrate pests, as well as weed infestations, were estimated to be as high as 30-35% in the field. Dinitrophenoles have had a huge impact on productivity. In its absence, post-harvest losses were estimated at 40-50%. It was estimated that without pesticides, world losses caused by pests in agriculture are in the order of US$ 75,000 million annually or about 35% of potential production. The difficulty in assessing the balance between the beneficial and detrimental effects of pesticides has led to what is termed as the "pesticide dilemma". Clearly, the balance can be shifted by minimizing the adverse effects through safe usage and maximizing the benefits by better and more efficient application methods (WHO, 2010).

3.4.13 Organotin Compounds

Organotin compounds are a stannaries' chemical compound made up of hydrocarbons. Information is lacking on actual pesticide handling practices that may explain how exposure occurs. Farmers regularly used organotin compounds pesticides and frequently mishandled them. An actual exposure has been witnessed globally including to animals that graze off in areas sprayed with these fungicides. It is important that farmers use it for its beneficial possibilities since it is effective and recommended for fungi elimination or control. Many cases of pesticide poisoning to occupational exposure in agriculture, it is important that you take precautionary measures when handling organotin compounds pesticides. Having an advisory roled out to farmers would help extension workers to advise farmers on better pesticide handling procedures to reduce the risk of the adverse effects of pesticides. Great caution must be taken to avoid common problems such as stomach upsets and eye irritations among others due to exposure of the organotin compounds (WHO, 2010).

3.4.14 Warfarines

It is an anticoagulant. From ancient times man has competed with other species of animals and plants for the available resources. In practicing agriculture, crop yields have often been diminished due to insects and rodent damage, plant disease and weed competition. In addition, man has been a host for a variety of disease-causing organisms vectored by pests. Thus, through the ages, there has been great interest in compounds and methods to control or eliminate pests and diseases. Warfarines have declined in tonnage is due to decreased usage of copper-based pesticides which were widely used in coffee. In addition, many currently used Warfarine are effective at low concentrations. Therefore, the decline in tonnage suggests an increase in the use of active ingredients which are effective at low concentrations and not a decline in total usage. Currently, substitutes of Warfarine include organophosphates, carbamates, synthetic pyrethroids and organochlorines as well as various classes of fungicides and nematocides.

3.4.15 Indanodiones

Indanodiones are organic compounds made of carbon, hydrogen and oxygen. Indanodiones is also known to cause severe diabetes and chronic neurological disorders. Indanodiones is an anticoagulant and increases hemorrhage during gestation. When administered as a drug, it is teratogenic in humans although there is controversy about this. However, compared to herbicides such as paraquat that is known to be acutely toxic on ingestion. In addition, a case occurred in which paraquat was found to be toxic on skin contact as well. Skin contamination from a leaking sprayer caused the death of one worker and reduced lung function in several others. Indanodiones, therefore, cause a wide range of effects including cancer, sterility, lung damage, diabetes, gross brain damage and bone necrosis. In the absence of surveillance, important human effects may be missed. It is important to take precautionary measures when using indanodiones to avoid harm or poisoning.

3.4.16 Bipyridyls

Bipyridyls are fungicides used to control fungi and are popular in large operations such as coffee plantations. Metal catalyzed reactions are involved in fungicide application in coffee is to spray the herbicide in the two meters inter-row for traditional cultivars and one-meter path. The use of fungicides in weeding coffee is comparatively cheaper than hand-weeding, but profitability is influenced by changes in coffee prices. The timing of bipyridyls application, followed by spot application of a contact herbicide is more economical in high rainfall areas than use contact herbicides alone or clean weeding. The use of low rates and volumes of recommended bipyridyls can be used to effectively control annual weeds in coffee. For better control of annual weeds, with- low rates, it has been recommended to control them at the 1 - 3 leaf stage. The low rates and volume used would be a saving to the farmer and an improvement in environmental safety. It is a control method commonly used to control pests.

3.4.17 Chlorophenoxy

This pesticide is used to control broad-leaved weeds. The need for comprehensive pesticide laws increased with the discovery of synthetic chlorophenoxy pesticides in the mid-20th century which are effective fungi control. Laws were needed to address the problems of safety to crops, to the users and to the environment. In the United States, the Federal Insecticide Act was superseded by the Federal Insecticide, Fungicide and Rodenticide Act of 1947. The fungicide chlorophenoxy is common sense it is very effective and does not have possible carcinogenicity. In all there are about 200 fully registered pesticides and a further 200 applications that are pending registration. Cash crops, pesticides play a major role in controlling pests and diseases and allow producers to secure markets for their produce. The use of chlorophenoxy is used extensively in the control of coffee berry disease allowed and thus common in many countries that produce coffee on large scale.

3.4.18 Glyphosate

Glyphosate kills broadleaf plants and grasses. In the last century, a fungal disease of American chestnut trees was controlled with glyphosate fungicides. Likewise, in many countries, the cypress aphid was partially controlled by spraying with systemic glyphosate. Furthermore, there are many reported cases of plant-pathogen resistance to fungicides. The glyphosate fungicide has been mistaken for soap and used to wash diapers for newborn infants. Therefore, specific effects are known or suspected are generally more informative. Glyphosate pesticide handling and usage were investigated among a group of three hundred horticultural farmers in an irrigation scheme in central many countries. With the help of a questionnaire covering farm size, literacy level, cropping practices, crop and household pesticide use, animal husbandry, pesticide storage, handling and labeling, disposal, foods consumed and health history; glyphosate has been found to be very effective in eliminating fungi. A wide range of crops was grown in the scheme and pest and disease problems were considered major constraints to production.

3.4.19 Acetanilides

Acetanilides are odorless and have a leaf-like appearance. These leaf crops have a high surface/volume ratio and thus have a greater concentration of residues than crops with a lower ratio. Acetanilides thus exposure from eating contaminated kale and cabbage was quite likely. Acetanilides have a low ADI of 0.005 mg/kg body weight. In the United States, registration has been pending since the 1970s. In many other countries it is only recommended as a soil treatment. Farmers and their families are likely to get exposed to carbosulfan on eating beans sprayed with the pesticide. Fanners re-entered sprayed fields to harvest without regard for the pre-harvest interval. They were, therefore, likely to contaminate themselves with sprayed foliage. Women workers were bound to re-enter treated fields at any time regardless of when the last spray was applied. Elsewhere, it has been found that harvesters may be exposed to acetanilides pesticides in this way, and in some cases get sick.

3.4.20 Triazines

Triazines contain nitrogen heterocycles and it is a commonly detected pesticide. This is not surprising since it is widely used for the treatment of cereals against insect pests in storage. Since it is normally mixed with grain, farmers may have the impression that it is safe and therefore handle it less carefully. Triazines are also very commonly used insecticide in the scheme and contamination is likely from storage and poor handling of either contaminated clothing or the pesticide itself. Triazines are also frequently sprayed in the field and their presence on tables may also be due to careless handling and poor storage practices. Triazines, however, are used as a household pesticide to control cockroaches and bedbugs and therefore its presence on the tables may be related to these usage practices. The presence of these pesticides inside the houses is undesirable. There are reported incidences of human intoxication with pesticides when people were exposed indoors (Levine, 2007).

3.4.21 Aluminum and Zinc Phosphide

Aluminum and zinc phosphide contains aluminous properties for poisoning insects and are sprayed on fields to harvest without regard for the pre-harvest interval. They are, therefore, likely to contaminate themselves with sprayed foliage. It has been found that harvesters may be exposed to pesticides in this way, and in some cases get sick. Aluminum and zinc phosphide are stored in the houses and only a small proportion is securely stored. This would facilitate the spreading of fumes from the chemicals throughout the house. Exposure contaminated indoor air can lead to illness. Different types of spray equipment lead to different levels of exposure. Water and other cleaning materials should be available in case of safety in storage ensure that unauthorized persons may not use or handle pesticides except in the presence of a trained and competent person. This is especially important for people who handle large scale farming.

3.4.22 Methyl Bromide

Methyl bromide is odorless and colorless and used in regulations of pests such as weed, insects, fungi, nematodes and rodents. The use of methyl bromide in many countries was restricted and no imports have been approved. An inventory of the types of pesticides used or stored by farmers was done and it was found that methyl bromide was not stored or used by the farmers. However, dicofol, an insecticide often contaminated with methyl bromide is used occasionally. It should be noted that organ chlorine pesticides have been reported in fetal tissues giving an indication that babies may be contaminated in utero and born with pesticide residue. In terms of the efficacy of spraying different formulations, it was found that less than 0.1% of the applied dose of foliar sprays may reach the actual target, the insect. A large proportion is suspended in air and could be a 'source of contamination. Methyl bromide intended for home and garden use may be supplied with tamper-proof caps that children would find difficult to open.

3.4.23 Ethylene Dibromide

In different countries, acute poisoning from ethylene dibromide pesticides is considered a common occurrence. Wettable powders have been found more hazardous than liquid concentrates during the mixing process. Conversely, the efficacy of spraying different formulations and found that less than 0.1% of the applied dose of foliar sprays may reach the actual target, the insect. A large proportion is suspended in air and can be a source of contamination.

Several incidents of abuse by ethylene dibromide pesticide companies have been compiled by the Pesticide Action Network (PAN). These include misleading advertisements, inadequate labeling and packaging, untrained sales people providing inappropriate and sometimes dangerous advice. Moreover, many third world countries do not have the expertise or resources to ensure that dangerous chemicals are safely used.

3.4.24 Diethyltoluamide

Diethyltoluamide is a chemical solution that is used in the synthesis of waxes, resins, dyes, gums and pharmaceuticals. For example, general repellents or systems to avert insect feeding on crops are potentially useful and safer than current chemical pesticides. In addition, pheromones, growth hormones, plant derived compounds, microbial systems and sterilization techniques are alternatives to conventional pesticides. Genetically engineered plants and animals that are incorporated with genes for pest control are expected to be produced in future and will play a wider role in crop protection. Integrated pest management (IPM) involves the use of multiple applications of diethyltoluamide in crop protection. The use of this approach gives chemicals a lower priority by requiring that they be used in smaller concentrations in combinations with other methods. Thus, improved agronomic practice, improved land use management, plant selection and ecological manipulations are aspects of integrated pest management that are likely to reduce the amounts and thus hazards associated with pesticide usage.

CHAPTER 4

Benefits and Risks Associated with The Use of Pesticides

CONTENTS

4.1 INTRODUCTION

Pesticides are both beneficial and risky at the same time. This chapter has very important information on these benefits and risks as discussed below.

4.2 BENEFITS

4.2.1 Crop Protection

Without pesticides, farmers could potentially lose half of their yield to diseases, weeds and vermin. As such, they help landowners to produce more crops on less land through prevention of bug and disease infestation, for increased output per hectare.

Upon planting, food crops typically compete with more than 30,000 different weed species, 3,000 worm varieties and 10,000 kinds of plant-eating pests. These threats not only stop upon harvesting but even afterward microbes, molds and rodents may cause destruction during storage. Insecticides can lengthen the lifespan of yields and avert post-harvest losses (Hokkanen & Lynch, 2003). Some common plant diseases that can be prevented by pesticides include:

Frogskin disorder: It's marked by the inability of roots to develop starch. Even though most crops don't show signs of this disorder in the above-ground sections, some affected foliage may show clues in the leaves.

Bean root rot syndrome: Many soil-borne fungal diseases are found throughout bean-growing zones of tropical regions, these can entirely wipe out harvests, especially when hostile environmental conditions continue after planting and during flowering. Pesticides can combat these maladies and their pathogens, some of which include; Fusarium wilt, pod rots, and yellowing (Hokkanen & Lynch, 2003).

Cassava mosaic disorder: Also known as CMD, it's caused by various species of germs, transferred by whiteflies and distributed through infected crop cuttings. The latest epidemic of severe CMD occurred in Uganda during the late 1980s, as of now the virus has invaded close to 12 states in Africa and still continues spreading.

Furthermore, pesticides minimize exposure to harmful micro-organisms that not only destroy crops but can also affect human health.

Figure 4.1: A farmer applies pesticide for crop protection.

Source: https://www.radiozdk.com/2018/10/law-enforcement-will-be-last-resort-in-use-of-toxic-chemicals/

4.2.2 Food Preservation

Pesticides also assist farmers to supply plenty of healthy, year-round meals, which are crucial for human health. They lower the costs of generating food and animal fodder, thereby making grains, fruits and vegetables more readily available. In fact, a U.S. based research discovered that without pesticides, harvests of most vegetables and fruits would decrease by around 50-90 percent. Insecticides are widely used in crop and livestock farming to increase production and guarantee sufficient food supply (Hokkanen & Lynch, 2003).

Balancing food production and pesticides: The only issue is balancing a consistent, top-quality food ration with the demand to protect consumers from needless exposure to chemicals. High amounts of pesticide residue in food products may also not be healthy for human usage. Moreover, experience shows that, even, a pesticide that's believed to be harmless to humans may still have a few undesirable effects.

A case in point is DDT, which due to its ecological persistence and capacity to gather inside bodily fat can be harmful. But DDT has so far been banned in many countries due to its potential dangers.

In many countries, state-sponsored produce-assessment programs have been set up to control the appropriate use of agricultural chemicals like pesticides. The ratio of pesticide deposit in food is affected by various factors including:

- Variety and brand of insecticide used
- The type of rain, wind or sunshine exposed on the plant, and
- The processing system that food undergoes, including storage duration, washing or peeling.

While pesticides can drastically increase crop yield and create better food quality if used in large amounts they can cause health problems.

4.2.3 Material Preservation

Residual pesticides are typically applied on the surface of artifacts to provide protection for extended periods of time. The overall period it lasts depends on various factors like type of surface sprayed such as brick or wood, surface condition and formulation of the insecticide-whether it's found in liquid or dust form.

A good example of residual pesticides is termiticides, which are very durable. These are usually applied in the ground under residences to prevent underground termites from attacking. In comparison, non-residual insecticides don't have any residual activity, but instead, work at the moment and are very short-lived. Aerosol sprays containing pyrethrum fall under this particular category, they are typically used for terminating flying and crawling bugs (Hokkanen & Lynch, 2003).

Additionally, pesticides can be used in cultural heritage sites like museums for preserving anthropological, natural science, and herbaria materials. Nevertheless, proper care should be taken to identify, test, and apply the correct pesticides in order to avoid contamination or sensitive cultural materials and heritage collections. With increased public involvement in gathering, handling, and submission of cultural artifacts to museums for storage, there's even greater need for the use of pesticides in preserving museum collections.

Among the qualities of good pesticides is that they should be toxic to bugs but safe for humans. Pesticides can lose their efficacy if the target pests become resistant. Therefore, occasionally samples of insects should be gathered and assessed for resistance. In case this is noticed, then another insecticide without cross-resistance can be used. Similarly, it must be safe to people and domestic animals alike. Nobody should be exposed to danger during or right after spraying the product (Hokkanen & Lynch, 2003).

4.2.4 Disease Control

Pesticides are commonly used to control the multiplying of germs and microbes. Among the most prevalent problems, however, is a predisposition to generalize and give recommendations for farmers who could be having different pest problems. Likewise, insecticides aren't just used to treat plant-based diseases but those found in animals as well. Some of which are as follows:

Vector-borne syndromes: These include infectious illnesses like rabies, West Nile virus and Lyme condition which are carried and distributed by vector-carrying creatures like mosquitoes, ticks, and rats. The environmental protection agency (EPA) recognizes various pesticides and repellents, which may be used to manage these vectors.

Asthma and allergic reactions: Everyday household pests like cockroaches are known to cause asthma and allergies. Apart from recommending certain pesticides to manage these pests, EPA also gives members of the public details on how they can safely use these items within their environments.

Bacterial contamination: Different forms of microorganisms, such as bacteria, germs, and protozoans, may trigger infections in hospitals, food processing centers and public health centers. However, there are antibacterial pesticides that can manage these microbes and help avoid the spread of microbial diseases.

Prions: Various proteins found in human cells, especially the central nervous structure (CNS) of people and animals may be found in abnormal, contagious forms known as 'prions.' They share many features with viruses, plus may cause terminal diseases to patients.

Pesticides can be mixed in different forms including; water-dispersible powder, enzyme concentrate or suspension mixture. For interior use, the water-dispersible residue is the most widely used formulation in the majority of countries (Hokkanen & Lynch, 2003).

4.3 RISKS

4.3.1 Toxic to Humans

Pesticides can poison plus, unfortunately, they don't just harm pests but sometimes even people and animals when used inappropriately. They are lethal, and excessive exposure often results in various health effects and illnesses. Exposure to insecticides can happen in several ways. Farmers and other laborers can get exposed to pesticides in farms through crop treatment, planting and grain storage. Plus rural dwellers staying next to farms may be affected by pesticide drift as well (Levine, 2007).

Exposure may also occur in professional forestry and local pest control services, via the cure of wood with commercial preservatives, treating boat hulls using anti-fouling agents, including the treatment of domestic animals with anti-parasitic preparations, such as sheep dip.

In modern metropolises, we are also exposed to insecticides through the treatment of amenities, like public parks, pavements and kids' playgrounds. Plus, many people purchase insecticides off-shelf for residential and garden use. Besides, pesticide residues contained in food may further put users at risk (Levine, 2007).

The symptoms to be worried about are:

Chronic toxicity: Some pesticides are acutely toxic, meaning they can result in harmful and long-lasting effects after uptake, inhalation or skin contact. Symptoms are manifest shortly following exposure, or may come up after around 48 hours and may include; allergic reaction, sore throat, extreme fatigue or loss of consciousness among other conditions.

Endocrine disruption: Endocrine disruptor is a compound that interrupts body hormones and internal balance. Generally, hormones are chemical connectors of the body, which are essential in controlling various functions, such as growth and reproduction.

4.3.2 Impact on the Environment and Ecosystems

Pesticides are toxic compounds designed for release into the ecosystem. Even though each is designed to extinguish a particular vermin, a huge percentage of insecticides reach a venue other than their original target. They easily pollute the air, soil and water when running off the agricultural fields, or when not well discarded after use.

Research shows more than 98% of insecticides and 95% of herbicides usually reach a location far from their primary target, primarily because they're spread across large field expanses. Consequently, runoff can disperse pesticides into marine environments, whereas wind can distribute them to far-off fields, grazing spaces, human habitations and unpopulated areas, possibly affecting other species. Problems may further arise from poor production methods, transport and storing practices. With the passage of time, continuous application causes pest resistance, whereas its negative effects on species may even enable the pest's reemergence (Levine, 2007).

Aquatic system: Pesticides are found in rain, underground water, rivers, lakes, streams and oceans where they can be dispersed from their original source. There are various ways that insecticides can enter the water system, such as soil leaching, accidental spillage, or carried as runoff. When pesticide concentrations surpass the allowable limits for safety in rivers and groundwater, fish and other marine life may be affected by the contaminated water. Moreover, the application of insecticides to water bodies may cause marine plants to perish, therefore reducing the water's levels of oxygen and chocking the fish. Recurrent exposure of some insecticides can further cause biological and behavioral alterations in fish, which may diminish in populations, abandon their nests, develop lessened immunity to infections, and so on.

Figure 4.2: Pesticides can pollute marine ecosystems where fish live.

Source: https://blogs.umass.edu/natsci397a-eross/fertilizers-are-doing-more-than-you-think/

4.4 CAUSE OF CONCERN

4.4.1 Multiple Chemicals

Pesticides are chemical compounds designed to eradicate pests that affect farm yield or carry infections such as malaria and typhus. On average, there are more than 15 different chemicals used to make these products, and they have different side effects on users including the following:

Endosulfan: It is a creamy-colored solid pesticide that looks like form flake crystals, plus smells akin to turpentine despite not being flammable. Endosulfan doesn't appear naturally in the ecosystem, but is rather manufactured for purposes of controlling pests on crops, and acting as a timber preservative.

Glyphosate: It's among the most common weedkillers found in farms, house gardens and lawns. Glyphosate products generally contain glyphosate ingredient, together with a blend of other components that help enhance the assimilation of glyphosate into the crop. Glyphosate-based products can easily be bought from insecticide stores. They can affect various organs in the body, particularly the respiratory and gastrointestinal systems (Levine, 2007).

Hexachlorobenzene: Until the mid-60s, Hexachlorobenzene was commonly applied as a pesticide for protecting the seeds sorghum, wheat, onions and various grains from fungal infections. Some manufacturers also used it to create fireworks, ammo, and artificial rubber. While currently it's prohibited for commercial use in America, other countries around the world still allow it. The compound comes in the form of a white-crystalline solid not quite soluble in water, it's often produced as a derivative of wood-preservation plants or when setting ablaze municipal waste. Hexachlorobenzene can affect the development of babies; trigger liver problems and cause reproductive challenges. Scientists also believe it's potentially carcinogenic to humans.

4.4.2 Multiple Sources of Exposure

Exposure to pesticides can have varied side-effects ranging from basic skin irritation to more adverse symptoms like headaches and nausea. Moreover, some organophosphate insecticides are known to trigger severe conditions like convulsions, unconsciousness, or even fatality.

In patients, pesticide toxicity can be categorized depending on the method of exposure, route of exposure, or the bodily system affected. Classically, any pesticide is more dangerous when ingested orally than if sniffed, and more noxious if inhaled compared to dermal exposure.

Poisons function by changing normal body structures or functions. Therefore, toxicity may occur in different ways just as there are bodily functions. Poisons may affect just a single organ, or one could experience generalized toxicity where several systems are affected all at once.

Moreover, some side effects are swiftly reversible and don't result in any long-lasting damage. Whereas others may trigger reversible damages, though full recovery may require longer periods. Also, poisons may lead to irreversible impairment, even if the contact isn't fatal. Toxicity is often categorized into different types, dependent on the amount of exposure one has to the poison, including the duration it takes for the harmful symptoms to manifest (Levine, 2007).

Acute toxicity is when an individual gets exposed to small doses of a pesticide, this is typically known as "acute" exposure. Additionally, if the exposure comes into contact with your skin, it's generally considered as acute dermal contact. On the other hand, acute oral is a single dose consumed by mouth, whereas acute inhalation is a single dose that is inhaled. Likewise, when there's repeated or constant pesticide exposure then this is referred to as "chronic" exposure.

4.4.3 Multiple Routes of Exposure

Exposure to pesticides can occur through different identified routes which are oral, dermal, and inhalation based on the variety and function of the pesticide. Some popular sources of pesticide exposure among the general public are as follows:

Dermal exposure: This refers to the absorption of a substance through the skin, which is the most widespread route by which users are affected by pesticides or other similar chemicals. Usually, dermal absorption occurs due to splashing, spilling, or drifting when the pesticide is mixed, loaded or disposed of after use. It could also arise from exposure coming from huge volumes of residue. The extent of dermal absorption risk depends on the overall dermal harmfulness of the insecticide, degree of exposure, method the pesticide is developed, and the section of the body affected. Overall, wettable residues, dust, and grainy pesticides aren't as easily absorbed across the skin, compared to other bodily tissues compared to other liquid formulations like emulsifiable concentrates, which have a high ratio of the poison within a rather small portion of solvent (Levine, 2007).

Oral exposure: If excess pesticide enters the patient's mouth, it could cause serious infections, injury, or even fatality. Pesticides may be eaten through negligence or taken by people who are intent on causing personal injury. The most common causes of mistaken oral exposure involve those whereby, pesticides have been relocated from their first container to another unmarked bottle or container, typically by somebody else apart from the individual who's ultimately poisoned. There are several cases where people, particularly kids, have gotten poisoned from accidentally consuming pesticides from a discarded soda bottle.

4.4.4 Multiple Effects

The particular way that insecticide poisoning affects humans, including the extent to which various sections of the populace can be affected aren't totally understood. Nevertheless, the signs and indicators of acute pesticide harm are well recorded. Symptoms are what someone feels plus they may be manifested to others (Levine, 2007).

For dermal exposure, the patient may experience skin redness, lesions, and hot, flaky skin which are the telltale signs. You must learn and always be watchful to these signs and symptoms, right from the onset of poisoning. If a case of poisoning is detected, it's advisable to immediately and entirely eliminate the origin of exposure. Through this, you can avoid additional exposure and reduce injury. Timely identification of the pesticide poisoning symptoms, including the immediate and total elimination of the origin of exposure could save someone's life. This is particularly evident if the individual is unconscious, or even unable to talk clearly.

Nevertheless, note that the symptoms of insecticide poisoning may also be triggered by other factors, including infectious illnesses or chemical exposure. Hence, it can't be presumed that since a person is within the proximity of pesticides, then the onset of some symptoms of poisoning occur due to pesticide contact. However, if any symptoms occur post- contact with insecticides, you must contact a doctor for help without any delay. In case of any doubt about the cause of your observed signs, take the safest route of action. Nobody will ever disapprove you for seeking medical help for a fellow colleague, who could have been poisoned even if it later turns out they hadn't been (Levine, 2007).

CHAPTER 5

How Pesticides Gets Into the Environment

CONTENTS

5.1 INTRODUCTION

The widespread application of pesticides by agriculturalists and the public provides several avenues in which pesticides get to the environment. Once in the surrounding pesticides will have varied effects to living organisms. Pesticide gets to the environment mainly after application or spills. Pesticides applied through spraying will move via air and end up in other parts of the environment such as soil and water. Pesticides applied directly to the farm soil may be washed off the soil towards surface water or may even percolate in the soil and end up in lower soil layers and consequently contaminate groundwater. Pesticides injected into the soil will end up just as the former into our surroundings. Another possibility is pesticides injected directly into water bodies for controlling weeds such pesticides will lead to build-up of pesticides in water bodies and will contribute to pesticides in air after evaporation.

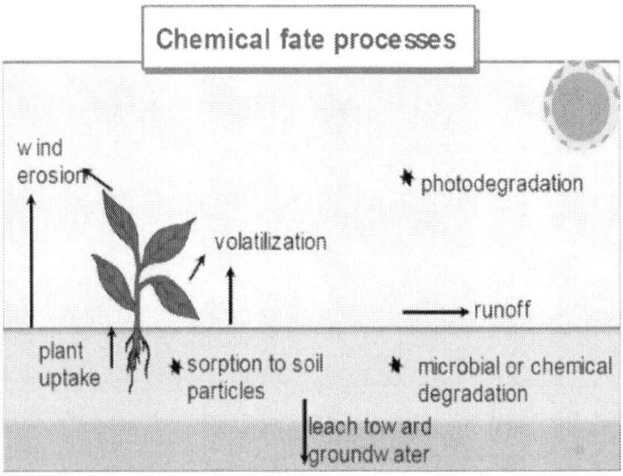

Figure 5.1: Processes of pesticide get into the environment

Source: https://www.slideshare.net/sundaershkalal/fate-of-pesticides-in-environment-or-environmental-polution-by-pesticides

Though the above possibilities exist, it is important to know that that not all pesticides will travel for long distances. Moreover, not all compounds will threaten our groundwater, surface water or the environment.

It is therefore important to note that once released to the environment pesticides will undergo a breakdown process known as degradation mainly caused by water, sunlight, chemicals or even microorganisms like bacteria. Degraded pesticides will thus lead to the formation of less dangerous products though some may produce more toxic compounds.

Secondly, some pesticides will resist degradation and remain in the original state in our environment for longer periods and therefore it is important to understand the types of pesticides and their effects on our environment so that one will be able to take appropriate action

Finally, it is important to reduce the unnecessary release of pesticides to the environment to reduce the exposure of pesticides to our environment. Moreover, integrated pesticide management should be adopted to help control the entry of pesticides into the environment through several pathways.

5.2 WIND DRIFT

Wind is an important weather element that affects how pesticides enters the environment. Wind causes spray drift. This is where a small droplet of spray drifts off target due to wind. Application practices and the nozzles of most sprayers also influence this. It is important to note that wind is a major cause of drifting. Though many believe that drift by wind is a major cause of pesticides and entry to the environment but most studies shows otherwise.

Figure 5.2: Pesticide wind drift.

Source: https://www.thermofisher.com/blog/proteomics/pesticides-in-ambient-air-a-novel-lc-hrms-protocol/

Wind drift depends on several factors i.e. application in orchard crops result in more critical drift in comparison with field crops because of sprayers involved and the sprayer orientation. In an orchard, spraying pesticides droplets is transported to the target or insects by air support and the spray direction is usually upwards (Racke, 2003).

The case is different in field crops because pesticides are directed downwards thus pesticide droplets are less exposed to wind, therefore, less wind drift of the pesticides. It is, therefore, true to argue that entry of pesticides to environment is more common in orchard farms compared to field crops. The types of nozzles are other important agents, which promote when drifting. Nozzle spacing should be usually checked to reduce wind drifting and projections from sprayers.

Wind speed is also a factor in the studies of pesticide drift. The higher the wind velocity- the higher is the wind drift on pesticides during application. To reduce pesticide entry into the environment one must consider wind speeds before pesticide application (Khan, 2016).

Additionally, wind drift is affected by an inappropriate wind direction during pesticide application. Wind direction drifts pesticides to nearby streams, livestock or even human population in the surrounding environment.

Additionally, air movement affects wind drifting. The distribution of pesticide droplets is largely influenced by their movement in the farms or in the surrounding area. Some fall slowly or remain suspended in air and later moved to other surroundings because of temperature inversion, which affects air movements.

Weather conditions such as relative humidity and temperature contribute to wind drift. For instance, low humidity or high temperatures result in faster evaporation of pesticide droplets, therefore, resulting in higher rate of drifting. Pesticide droplets will then be carried and introduced to the surrounding environments by wind (Khan, 2016).

In conclusion, though pesticides drifting cannot be eliminated the following measures can be taken to help reduce pesticide drift to reasonable limits to avoid danger to the environment;

- Apply low and non-volatile pesticides.
- Adhere to labels and instructions especially those related to the environmental risk.
- Use spray additives recommended by the manufacturer to reduce pesticide drift.
- Avoid high spray boom pressure during spraying to avoid creating fine droplets that made be drifted by the wind.

- Always use drift reduction nozzles, which will produce larger droplets, which cannot be carried easily by wind.
- Do not spray when temperatures are high or the wind is blowing towards unintended targets such as dwellings, water sources or livestock.
- Finally, use lower application speeds to reduce pesticide drifting.

5.3 EVAPORATION

Evaporation is a process where a liquid molecule becomes gas. Evaporation of pesticides is all about energy of individual molecules. Some of the pesticides sprayed on the field may leave from soil or from crop and surface to the air. The amount of evaporated pesticides will, therefore, disappear into the atmosphere through evaporation. Pesticide residues have been found in rain- a risk that endangers the environment. This is directly related to application techniques and other natural conditions and processes such as climate and soil properties.

After evaporation of pesticide particles, they are transported to the surrounding environment through air and wind drift and some are carried back to soil or water bodies by rainfall.

Several factors affect the rate of evaporation of pesticides. These factors include the temperature, humidity surface area and wind speeds. Temperature affects the rate of evaporation. Pesticides evaporate more in high temperatures, therefore, all pesticides should be kept in lower temperatures to avoid evaporation of pesticide particles. The pesticide must also be sprayed during conditions of moderately lower temperatures to avoid high chances of evaporation into the air. This is to avoid excess evaporation of pesticides into the environment. Secondly, the amount of water vapor present in the air called humidity affects the rate of evaporation of pesticides. The higher the humidity the lower is the rate of evaporation. Most pesticides evaporate faster during conditions of low humidity. Moreover, pesticides should, therefore, be kept where the room humidity is higher. Additionally, the rate of evaporation increases with an increase in surface area. Evaporation is affected by the surface area of the pesticides. Thus pesticide particles have to be smaller because this is the best way to reduce their evaporation to the environment i.e. the smaller the surface area the slower is the evaporation rate while the bigger the surface area of pesticide particles the faster is the rate (Racke, 2003).

Wind speed is another important factor, which affects evaporation of pesticide particles. Most pesticides spread faster to the environment when there is an increase in wind speed. All the pesticides used in farms evaporate faster with an increase in wind speeds. This climatic condition should be checked when applying pesticides to reduce evaporation of pesticides to the environment. This will reduce the chances of pesticide spreading too much further in the environment.

In conclusion, it is important to note that evaporation rates of pesticides differ from one pesticide to another. The loss of pesticides through evaporation from crop surfaces is much higher than from soil surfaces due to lower absorption capacity of plants. Moreover, pesticides with high vapor pressure are more likely to evaporate and escape into the atmosphere. It is important to note that when applying pesticides, one must select the most effective and environmentally friendly pesticides. This will help reduce the danger to living organisms depending on the surrounding environment for survival. Pesticide users should select pesticides that have low evaporation rates to reduce the loss of pesticides to evaporation.

5.4 SURFACE RUNOFF

Surface runoff is the movement of water and any contaminants across the soil surface. Runoff occurs naturally when irrigation, rain, snowmelt add water to the surface water. Running off to the rivers and lakes move pesticides particles with them.

Rainfall affects run off because pesticides move faster during heavy rains soon after spraying pesticides on unsaturated soil. Secondly, the slope of the area contributes to run off simply because on slopes the more likely water will run off the side. The vegetation cover effects runoff when cropping system leaves crop residue that slows movement of water and we will keep the pesticides on the site. One should be able to reduce control of the target to avoid environmental damage. Other usages can disrupt the natural polymers in the ecosystem generally pesticides are toxic to the environment.

Figure 5.3: Pesticide carried by runoff.

Source: https://www.eawag.ch/en/news-agenda/news-portal/news-archive/archive-detail/news/fluesse-sauber-halten-ist-vorsorge-fuers-trinkwasser/

Water-soluble pesticides are easily transported from one area to the other through run off. This leads to contamination of groundwater or streams once they dissolve. They cause damage to aquatic animals and plants. While fat-soluble pesticides are absorbed in insects and even fish. Soil moisture content is a significant factor that determines how much runoff occurs. Moist soils are more prone to run off that lead to losses of pesticides compared to dry soils. The soils that contain clay are prone to runoff compared to sandy soils. Manufacturers must assess the properties of their products related to solubility absorption and persistence and state the appropriate precautions on the label (Racke, 2003).

Highly soluble pesticides are more likely to be carried by running water and washed off -site. Most of the pesticides have poor water solubility. They will stay on the treated area for many years and can only be moved from that place when eroded by wind or other agents of erosion. The persistence of the pesticides in a treated area plays another role in surface runoff. The faster the pesticides are absorbed then lesser pesticides will be available for runoff. Some pesticide design formulations are very dangerous because they stay on -site and cannot be moved by rainfall or water runoff.

Pesticide particles may affect non-target species because pesticides used to kill fungi and pest reach their habitats by runoff. Runoff carries pesticides into the aquatic environments where else wind carries them to other Fields, grazing areas, human settlements and undeveloped areas. It is therefore important for manufacturers to label pesticides appropriately. Pesticides prone to run off be banned and those, which are more toxic after the runoff, should be avoided because of the danger to the environment. In conclusion, pesticide runoff is a serious environmental concern because of its negative impact on water supplies. This results in less growth and oxygen depletion in lakes and streams.

5.5 LEACHING OF PESTICIDES

Leaching is the movement of pesticide particles especially water-soluble pesticides, which are carried by water downwards through the soil. Generally, most pesticides absorb soil particles. Groundwater does not dilute contaminants that reach them. These pesticides may take many years to be flushed off the groundwater due the cold temperatures, lack of sunlight and low oxygen level levels in groundwater, which affects the breakdown of pesticides. Many soil characteristics affect leaching such as organic matter, soil texture, and structure and soil water content. First, the organic matter content in the soil is the most important soil property that affects pesticide breakdown by microorganisms and once broken down particles will, therefore, leach downwards and end up in groundwater. In other cases, degraded pesticides will nourish microorganisms and reduce the chances of leach of pesticides into groundwater. The organic matter can be increased by adding manure and growing cover crops (Racke, 2003).

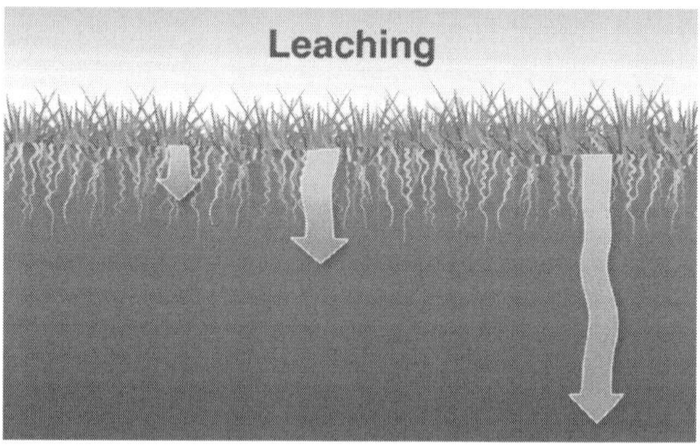

Figure 5.4: Pesticide leaching.

Source: http://nysgolfbmp.cals.cornell.edu/leaching/

Secondly, the soil texture affects the movement of water in the soil because the texture of the soil and soil particles with large pores allows water to move pesticides carried by water through the texture of soil. These way pesticides are more likely to reach and contaminate groundwater. Poorly textured soils, on the other hand, have low permeability and absorb more pesticides from the water this slows movement of chemicals and help improve absorption of pesticides in solid particles thus reducing groundwater contamination.

There are site characteristics that affect leaching; this includes the depth of groundwater, the type of bedrock and the slope. The depth of groundwater is the depth of the soil, which separates the water from the earth's surface. This soil will protect the groundwater by allowing pesticides absorption and evaporation. When the water table is high or close to there are surface, groundwater is highly protected from contamination. The type of bedrock as an avenue affects the floor of runoff water and groundwater. The bedrock is the foundation layer beneath the soil or rock fragments. Limestone bedrock has larger channels from the groundwater to the surface. This allows water to pass through quickly. The limestone allows quickly passage of water and reducing the chances for pesticide degradation occurrence. Limestone also dissolved in water (Racke, 2003).

The slope of the area also speeds the movement of water across the surface. The steep slopes allow faster surface runoff and reduce the passage of water into the ground in valleys and flat areas. Water flows slowly for it to be absorbed in the soil. In conclusion, leaching is more of an environmental concern. There are pesticide pathways in the soil through which the pesticides are leached into the water down through the soil. They can badly affect the environment making it dangerous for all living organisms (Racke, 2003).

5.6 WORMHOLES, CRACKS AND ROOT CHANNELS

Natural and artificial pathways provide avenues for pesticides to be introduced underground to the environment. The natural pathways include wormholes, root channels, cracks and soil factors such as swelling and cracks. They also include insect that burrows and micropores. Artificial pathways, on the other hand, are dry; wells and utility lines, which end up transporting pesticides to the underground environment.

Several factors affect these artificial and natural pathways underground to allow the movement of pesticides into the environment. First, the length of time these pesticides remained in the soil without being broken down depends on how accessible a pesticide is available for breakdown by microorganisms or other forms of breakdown. Some pesticides survive for a long time in the smallest soil cavities simply because microorganisms are large to enter these small pores and hence these pesticides are protected from breakdown. Mechanical problems common to soil penetration by pesticides into root channels and warm holes. The hazard of pesticides and reaching to the groundwater through roots channels can be passed to the subsoil and remove pesticides (Racke, 2003).

Physiochemical parameters of soil are necessary to determine the behavior of pesticides in the environment. Micro pores such as cracks road channels and warm holes are shortcuts to create caps and consequently underground water. The persistence of pesticides in the soil is related to its chemical structure which determines the downward movement through route channels or cracks and wormholes. Moreover, loosely packed soil particles allow speedy movement of water through the soil once mixed with soluble pesticides. Water and pesticide particles when in warm holes and other micropores and are evenly distributed throughout the farm or may even flow downward leading to the contamination of groundwater.

In conclusion, one should be able to read pesticide labels to make sure they are using the product accordingly to minimize the risk of environmental damage. Use pesticides specifically designed for pests you are controlling rather than a broad spectrum that is likely to affect the organism and their surroundings at large. Disposal of pesticides and pesticide contaminants should be done properly to avoid consequences of contamination to reduce the chances of pesticides ending up in warm wall cracks and route channels and consequently to our environment (Racke, 2003).

5.7 ADSORPTION

Farmers and other companies control pests by the use of pesticides and therefore efforts must be made to reduce the entry of these toxic chemicals into the environment. Several studies have been carried out to highlight the absorption of organic compounds in soils particularly on pesticides. There is much debate on the mechanisms and approaches to explain and predict adsorption and soil properties of pesticides and those of the soils.

There are three factors, which do not necessarily affect the absorption of neutral compounds that have an influence on the adsorption of pesticides: soil factors and soil pH play and oxide components. Adsorption of pesticides may be the spontaneous deposition of a solute into a solid sorbent due to interaction between fields of force of the solute and the molecules of the absorbent pesticides. Absorption process including chlorophenols and acetic acids into the soils and sediments are influenced by several factors such as physical and chemical properties of the soil (Racke, 2003).

Several studies reveal that soil pH results in low adsorption rates while organic soils tend to adsorb more pesticides. It is thus crucial to note that areas with lower soil pH and with clay soils will have no channels of pesticide absorption, therefore, less environmental contamination.

Soil temperature plays a critical role in absorption of pesticides. Absorption is generally exothermic and any change in temperature affects pesticide solute and water solute interactions and consequently may increase or reduce the rate of absorption of pesticides. Moreover, organic matter in the soil has crucial effect on adsorption of pesticides. Clearly, soil of higher organic matter adsorbs more pesticides. For instance, topsoil will adsorb pesticides more because they contain high levels of organic matter (Racke, 2003).

In conclusion, it is important for agriculturalists to understand the adsorption process keeping in mind that sorption plays a critical role in introducing pesticides to the environment. It is therefore important for farmers and other pesticide users to adhere to the following in an attempt to reduce the entry of pesticides into our environment;

- Pesticides with high the rate of absorption should be used to help reduce environmental risk.
- Soils with low organic matter should be avoided because they will not absorb pesticides.
- Reduce pesticide usage on bare and sloppy farms to reduce their chances of reaching nearby streams or plant and animal habitats.

Further studies and research on adoption will help environmentalists reduce the danger of pesticides to the environment and understand their behavior after adsorption. These particles will end up in groundwater and surface water in different parts of the world. Pesticides cause hydrophobic partitioning ionic exchange and water Bridging and formation of bound residues, all which may remain in the environment and will lead to danger to human and animal habitats (Racke, 2003).

5.8 VOLATILIZATION

Pesticide volatilization involves the movement of pesticides vapor through the air. This movement will cause the surrounding environment and particularly living non-living organisms to be exposed to pesticides.

It was originally believed that pesticides enter the atmosphere during application through spray drift but evidence from recent studies shows otherwise. This has been supported by the presence of pesticide samples in seawater and consequently being detected in sea fish, sea mammals and plants. After volatilization pesticides are transported by the atmosphere into areas far away from the original source where they may be introduced to the environment from different forms and ways. Volatilization reduces the quantities of pesticides available to control plant pests and have the potential for contamination of groundwater or general environment. Most pesticides are toxic and carcinogenic and therefore their volatilization will increase the risk of illness especially to living organisms. This will result in cases of cancer and other related illness (Racke, 2003).

The major sources of pesticides available for volatilization are important in the understanding of the key process of volatilization of pesticides. First pesticide drifting during spraying is primarily the open- source of pesticide volatilization. During application through spraying wind will drift pesticide droplets from plants and their surfaces or even during transport from nozzles to the plant. This way the light pesticide particles will be carried through volatilization leading to contamination of air and other parts of the environment. Secondly, pesticides in the soil will be dissolved from the soil solids into the soil solutions and finally into the soil air. Once on the surface the soil pesticide will be moved into the atmosphere by air carrying with them pesticide particles (Matthews, 2008).

Volatilization from soil can be affected by a number of factors which may include the properties of specific pesticides such as the vapor pressure and pesticide water solubility other crucial factors are soil factors such as soil moisture, organic matter, soil temperature and transport abilities of specific soil type.

Additionally, pesticide volatilization from plant surfaces is key in the understanding of our pesticides reaching our environment. Volatilization from plant surfaces is more compared to that from soils. Volatility from plants is influenced by vapor pressure, the absorption of the chemicals by leaf services and rainy weather that may wash off pesticide residues off the plant or other surfaces. Finally, environmental conditions such as temperature and humidity play a critical role in affecting volatilization process of pesticides from plants. The high temperature will lead to higher levels of vapor in the atmosphere and therefore high rates of volatilization of pesticides. It is therefore useful for agriculturalists and all pesticide users to consider all these underlying factors in an attempt to unravel the fate of pesticides in our environment and habitats. The study of all the factors will be useful in government institutions and agriculture to be used in the management of pesticides. Volatilization occurs over a long period (Matthews, 2008).

5.9 BIOLOGICAL BREAKDOWN

Pesticides help in controlling and preventing diseases and pest on crops and livestock. Sometimes pesticides and their residues have serious environmental hazards. Most pesticides that reach the environment through microbial degradation in the soil environment are worth to study so as to understand the microbial organisms involved and the mechanisms of microbial degradation by bacteria.

Furthermore, it is necessary to understand degradable pesticides in our natural environment, it is important also to know that pesticides factors such as molecular weight, structure, type of substituents and location have a great impact on biodegradation. In general, low-molecular-weight polymer compounds are more broken down by microorganisms. Microbial degradation is a slow process that involves using pesticides as microbial nutrients and leading to decomposition into carbon iv oxide and water through and symbiotic reactions. This process will lead to release of carbon iv oxide and water which are less harmful to the environment (Hokkanen & Lynch, 2003).

Soil microbes may degrade pesticides into less harmful compounds through mineralization. This is usually achieved when organic compounds in pesticides are converted to inorganic compounds by the actions of soil microbes. Mineralization is important in the biological breakdown of pesticides into less dangerous or non-toxic inorganic compounds.

The biological breakdown is affected by several factors; the type of pesticide, microbial species and their metabolic activities. Different microbial species have different reactions to pesticide compounds. Compounds will induce microbial organisms to produce enzymes used in the breakdown of pesticides (Hokkanen & Lynch, 2003).

Additionally, some organisms have a fast metabolism in degrading pesticide compounds while others do not have thus they will degrade pesticide companies slower or faster depending on the pesticide compounds.

Environmental factors such as temperature and humidity play a critical role in the biological breakdown of pesticides. Temperature and humidity affect the growth and reproduction of bacteria and when conducive environments exist more bacteria will reproduce and breakdown pesticide compounds. Many microbial organisms involved in the degradation of pesticides are found in sewage or soil and they include algae, fungi, and bacteria.

Recently, microbial degradation of pesticides is considered much better than other degradation processes like chemical breakdown. It is believed to be economical and produces good results. Biological breakdown, however, may leave or produce more toxic compounds in the human habitat. Moreover, some pesticide compounds are resistant to biological breakdown, and therefore pesticides in their enter and harm the environment (Hokkanen & Lynch, 2003).

In conclusion, although using pesticides is a necessity in agricultural practices, environmental problems have increased over the years and therefore this threat to the environment needs to be controlled through proper pesticide management. The entry of contaminated products to human and animal environments is causing harm due to bioaccumulation. Despite the fact that some pesticide compounds may be degraded slowly by microbial populations the process is low thus leaving toxic pesticide compounds that are not yet biodegraded to cause hazards to both human and plant health.

Moreover, farmers and pesticide users should adopt practices that ensure safe biological degradation of pesticides. Environmentalists should determine appropriate action with weather and soil properties that affect the process of pesticide biological breakdown.

Finally, environmentalists should identify pesticide degrading microbial strains and take steps to inform pesticide users on the application of such biological degradation to reduce the risk of pesticide entry in habitats.

5.10 CHEMICAL BREAKDOWN

Chemical breakdown is another pathway, which allows pesticides to end up into the environment. Chemical degradation of pesticides is the breaking down of pesticide molecules into smaller particles. Applying pesticides plays, a bigger role in chemical breakdown once applied pesticides reach the soil and contribute to chemical breakdown. Pesticide chemical breakdown is high when applied over a long which leading to toxicity.

Several studies show that pesticides used on pests failed to reach a target pest but end up contaminating soils, water and the air. Moreover, pesticides on reaching soil may change the soil chemistry while altering soil fertility (Capri & Karpouzas, 2011).

Chemical degradation caused by pesticides affects soil microbes and microbial populations. After break down of a pesticide, molecules become toxic killing microorganisms or altering their reproduction. Furthermore, pesticides suppress symbiotic nitrogen fixation, which consequently reduces crop production.

Hydrolysis is another type of chemical breakdown of pesticides, which involves the reaction of pesticide molecules with water. Hydrolysis involves catalysis by protons, hydroxides, or even inorganic ions like phosphate ions found in aquatic environments. Hydrolysis is useful in studying the toxicological impact on aquatic life and where sediment dwellers as are found. Hydrolysis is common in groundwater and surface water and the results in normally contamination, which makes water dangerous for humans, animals and aquatic life.

Chemical breakdown also occurs in the atmosphere where pesticide molecules are degraded due to reaction with air compounds. Pesticides carried by wind after being applied enters the atmosphere and end up combined with dust particles. This will lead to the reaction with atmospheric oxidants and end up being degraded. This process is called heterogeneous atmospheric degradation (Capri & DKarpouzas, 2011).

In conclusion, it is important to highlight the environmental significance of each type of chemical breakdown of pesticide compounds this is to assist in making well-informed decisions and proper management of pesticides by both farmers and environmentalists;

- Most pesticides are less polar and therefore will accumulate more in soil or living organisms than in water.
- Pesticides, which resist hydrolysis, will be degraded through other pathways.

Furthermore, it is useful to know that after the breakdown of pesticide particles through chemical breakdown some particles are rendered harmless while others become more toxic and thus the following factors need serious consideration;

- Pesticides after breakdown are transformed into either active or inactive metabolites.
- Fat-soluble pesticides will end up being stored in adipose tissues.
- Pesticides in living organisms may be eliminated through urine.
- Toxic kinetics of pesticides depends on chemical characteristics and formulation.

Pesticides producers should consider chemical breakdown during the production of pesticides. All stakeholders should take steps to reduce exposure to pesticides or pesticide products after chemical breakdown so as to help reduce any further exposure to the environment.

5.11 PHOTOCHEMICAL BREAKDOWN

Photochemical breakdown of pesticides also known as photolysis involves the breakdown of pesticide particles by sunlight. Pesticide particles that are lost to the atmosphere through volatilization end up being broken down by sunlight in photochemical reactions.

Products produced by photochemical reactions may be more harmful or less toxic than the current chemical composition of the pesticides. Photochemical breakdown may take in water especially if enough sunlight wavelength is present (Damalas & Eleftherohorinos, 2011).

Sunlight moving in wave-like manner produces bundles of energy called photons. Sunlight emits photons of different wavelengths. Pesticide particles can absorb photon energy leading to chemical reaction, which breaks molecules in pesticides.

Pesticide particles are broken by sunlight either directly or indirectly. A pesticide molecule may be broken directly especially those with weak bonds. Indirect breakdown occurs when sunlight breaks chemical bonds in the air, which in turn react with pesticide molecules, and breaking them in a process known as photolysis (Damalas & Eleftherohorinos, 2011).

Several factors affect the process of photochemical breakdown. First, time of the day is crucial in this photochemical breakdown. The rate of photochemical breakdown is determined by available sunlight and its wavelength. This is because photochemical breakdown depends on amount of photons that can react with pesticide molecules for instance during mid-day sunlight causes photochemical breakdown. Moreover, photochemical breakdown will increase with the temperature of sunlight and the surrounding thus maximum photochemical breakdown.

Weather plays an important role in photochemical reactions. Pesticides will not be broken in cloudy weather. This is because clouds will prevent sunlight from reacting with pesticide particles and making photochemical reactions slow or ineffective.

Thirdly, the type of pesticide particulates influences photochemical breakdown. Some pesticides are not degraded by light because the sunlight wavelength required is not available. For instance, pesticides absorbed by dust or other solid particles in the air or water will change their bonding strength and therefore will require higher wavelength to break. Moreover, the type of particles in the air or water may prevent photolysis. Particles that scatter light waves prevent photons from degrading pesticide particles. Finally, photochemical reaction involving water bodies depends on water depth, which greatly affects the photochemical breakdown. For instance, the depth of the water which is mixed with pesticide particles will reduce photochemical breakdown due to reduced light available for photochemical reaction.

In conclusion, it is crucial to understanding the process of photochemical reaction because it is an important pathway through which pesticides enter the environment. Products formed because of the photochemical breakdown may turn active toxic or less harmful products than the original pesticides. After the degradation by solar, pesticide products may be removed through precipitation out of air and return to the earth's surface through rainfall or wind movement. In the case of photochemical breakdown in water bodies, reaction products end up in aquatic environments and organisms. It is therefore important for pesticide users to understand photochemistry of pesticides in rigid media and plant metabolism. Products of photochemical reaction should be analyzed and studied closely by agriculturalists and environmentalists to determine their toxic levels in the environment and to reduce emission of sunlight on stored or spilled pesticides. All these are aimed at securing a cleaner and safer environment (Damalas & Eleftherohorinos, 2011).

CHAPTER 6

Children Exposure to Pesticides

CONTENTS

6.1 INTRODUCTION

Different people are predisposed to pesticides differently. Among them are children, who are at great danger of pesticide poisoning. This chapter provides important information concerning children's exposure to different forms of pesticides in different settings.

6.2 CHILDREN EXPOSURES AT HOME, SCHOOL, DAY-CARE AND INSTITUTIONS

Kids are at greater risk of developing health problems from pesticide exposure compared to adults since their inner organs are still maturing. They may come into touch with pesticides kept or spread in their residences, yards, daycare centers, learning centers, parks, or even pets' skins. Youngsters, as parents know, often put their hands inside the mouth. Plus, they usually play on floors, lawns, or other spaces that could contain pesticides. Since pesticides are found in multiple venues in the environment, the kid's amount of exposure may rapidly add up (Philip & Landrigan, 2018).

Pediatricians have comprehended for years that kids are more exposed to harmful chemicals than adults. This increased contact in children has been associated with attention and learning challenges, including cancer.

6.3 MOST COMMON SOURCES OF CHILDREN EXPOSURE

Neonatal: According to the AAP (American Academy of Pediatrics), pesticide residues found on food are the most common route of infant exposure. When a fetus gets exposed to some types of chemicals at certain periods, the brain undergoes developmental problems, including some organs which may fail to take shape, plus some of these side-effects are irreversible in youngsters. Pesticide exposure in times of pregnancy could cause a heightened risk of childhood defects, decreased birth weight, and even fetal loss.

At the household or daycare center: If insecticides are used at home, in the yard or gardens where a child is discovering the world around them, then exposure may be certain.

At schools and parks: Similarly, the use of toxic compounds to manage pests in schools, or at playing fields could make the learning environment less accommodating for developing bodies and brains.

6.3.1 Indoor and Outdoor Application

Pesticides are crucial in fighting against plant diseases, plus they are widely used for bug control in gardening, residences, and soil treatment. Insecticides, when not properly handled, may upset human body structure, or organ functions, in both men, women and children (Philip & Landrigan, 2018).s

Indoors: Some common factors that cause illness indoors are excess pesticide application, pesticide-treated bedcovers, and inadequate notification of pesticide use. Most of insecticide exposures are caused by pyrethroids, pyrethrins, and other toxic ingredients found in pesticides. They are responsible for causing health-related neurologic symptoms like headaches, drowsiness, gastrointestinal problems and respiratory issues such as upper respiratory tract aching, irritation and dyspnea, including other symptoms like uneasiness and vomiting. Some insecticides may also disrupt the hormonal balance that's crucial for proper bodily functioning. Besides, homeowners should take into consideration that dose, timing, and period of exposure are crucial to the probability for a pesticide to trigger harmful effects (Philip & Landrigan, 2018).

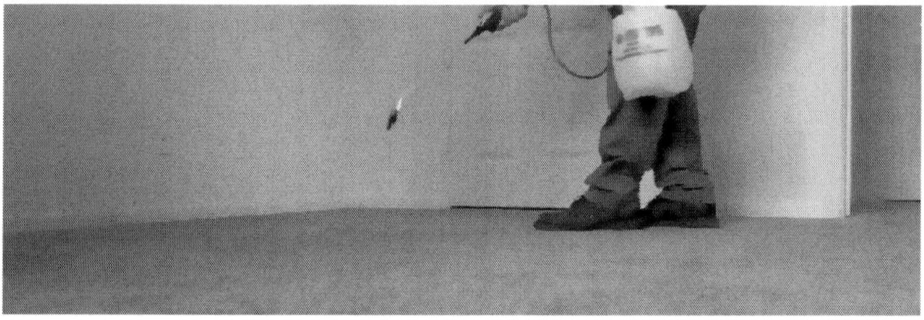

Figure 6.1: Families risk pesticide contamination through indoor applications.

Source: https://www.solutionsstores.com/carpet-beetle

Outdoors: When used outdoors, pesticides are applied to eradicate or control pests that include bugs and rodents, as well as bacteria, fungi, or other organisms detected in your compound. Pesticides used on lawns or gardens may further drift into the house, where it could be found in the air. Such exposure may cause irritation to the eyes, nose, or throat; also damaging the central nervous system (CNS) and kidney, or increased cancer risk. Rural kids are particularly at greater risk, since they may be faced with pesticides coming from nearby farms into their residences or onto schoolyards, polluting the water they consume every day.

6.3.2 Pesticide Residues

Dust, soil, furniture, carpets, toys, and food: Children are generally believed to come into contact with more pesticides through toys, carpets, soil and other surfaces they frequently interact with. Residential pesticides are spread when treating the household or pets for bugs or the lawn, patio, or yard garden for weeds. Toxic compounds stay behind in the dust, soil, furnishings, upholstery, toys or even food that's left exposed. Studies further reveal that residential pesticide usage emerged primarily from discoveries of higher amounts, detected in rural/ agricultural homes reporting the frequent application of the product to exterminate weeds and fleas, compared to households that didn't treat for these particular pests. Those living in close proximity to plants or orchards where insecticides are often applied tend to be at a higher risk of contamination. Likewise, approximately 80 to 90% of homes in both township and rural areas are reported to use pesticides in household and lawn pest treatments (Philip & Landrigan, 2018).

The most commonly identified chemicals are pyrethroid pesticide permethrin, with a detection rate of 100%, followed by herbicide 2,4-D at 98%, and the catalyst piperonyl butoxide at 97%. Likewise, the organophosphate pesticides diazinon and chlorpyrifos, including the herbicide mecoprop had immense detection rates. These compounds are usually applied to deal with everyday vermin like; fleas/ticks, ants/cockroaches, flying insects and garden insects.

Scientists are still trying to identify the correlation between insecticides and human disease, though research shows that exposure to various pesticides can cause trigger certain cancers, such as non-Hodgkin lymphoma or leukemia. Similarly, pesticides have been tied to the risk of autism in babies, including attention-deficit hyperactive disorder (ADHD) in younger children, as well as teens.

Playgrounds, playing fields, lawns, and gardens: In a study conducted in Italy, it was found that pesticides covered around half of all the 71 public playgrounds sampled in the tested. A total of twelve different pesticides were discovered, plus around 90% of the compounds identified were 'hormone active,' meaning they contained endocrine- disrupting elements. These kids' playgrounds were contaminated with insecticides spreading from surrounding apple and wine fields.

Such public spots are usually exposed to pesticides drifting from surrounding agricultural fields. Moreover, the level of contamination of lawns, playing fields, gardens and playgrounds depended on their distance from, farming fields in the neighboring landscape, including from wind and downpour intensity. Pesticides applied on farms to ward off pestiferous bugs, weeds or infectious fungi may easily be dispersed, floating away with the winds or carried by rain. While farm products are frequently monitored for signs of insecticides residue, there's usually no monitoring done in public zones or private gardens. Children in particular, are worse affected since these compounds can disrupt early development, that's especially a sensitive growth phase. Exposure to these endocrine disruptors may also have lasting harmful effects, and may possibly trigger cancer, reduced brain functions, weight gain and diabetes. To prevent pesticide contamination, researchers estimate that the space between playgrounds and farmyards should be no less than 100 m. However, with strong wind conditions insecticides could disperse even over 300m away. Likewise, higher residue proportions are probably based on dispersal conditions. Authorities should ensure preventive measures are put in place so that playgrounds don't become a health risk to youngsters due to pesticide drift (Philip & Landrigan, 2018).

Wood preservatives in play structures (e.g. PCP: pentachlorophenol): There are various types of timber preservatives found in play structures, which include waterborne, oil-borne and fumigants. Their effectiveness in the preservation of these products depends on factors like the formulation chosen, ratio of sapwood to heartwood, application method and the way it's distributed on the product. Some common examples of wood preservatives found in play -areas include:

- Pentachlorophenol: It was first developed in the 1930s, being an artificial organic wood preservative. The compound is very effective in combating fungi and insects. At some point, it was the single most extensively used preservative, though concerns regarding the existence of dioxin contaminants caused a number of states to impose sanctions against its use. In appearance, Penta is crystalline and is usually immersed in light-petroleum liquid first before use. The timber products commonly treated with pentachlorophenol are; construction timber, utility poles, fencing posts and crossarms, some of which are found at the playground.

- Chromated copper arsenate: Also known as CCA, this is commonly used in fencing posts around the playground. It's a coal tar distillate which enhances the anti-weathering properties of wood, offers protection from pests and fungi, and provides insolubility in water.
- Copper naphthenate: It's another common timber preservative that's typically diluted with mineral spirits or diesel fuel, the compound is applied by brush, rolling or immersion onto the wood surface. It works by preventing timber decay, termites, other timber attacking vermin, and mildew.

These compounds can be toxic when they come into contact with children; they often leave behind a pungent odor that's harmful and irritating to humans, plants and animals.

Long -range transport of POPs (e.g. DDT): POPs or persistent-organic-pollutants are organic elements of anthropogenic origin and include products such as; hexachlorocyclohexanes, pentachlorophenol and DDT. They have properties that repel photolytic, organic and chemical degradation, thereby causing bioaccumulation within the food chain. Additionally, POPs can easily be carried over vast distances in the environment, resulting in vast distribution across the terrain, including places where they've not been used yet. As a result of their toxic properties, they can be dangerous to people and the environment at large. In more recent times, the global community has asked for urgent universal action to minimize and abolish the usage of POPs, as well as identify possible risk factors to people's health and the surroundings at large. Some common types of POPs are aldrin, dieldrin, endrin, chlordane, heptachlor, chlordecone, DDT, and mirex among others (Philip & Landrigan, 2018).

Generally, the physical and biochemical properties of this element suggest slow evaporation in the air, but mostly it spreads by water and sometimes even with soil particles. The movement of PCP in the ground is also determined by soil acidity, since the synthesis of PCP in the water is a significant process that's reliant on the pH. Its ionized form also tends to be more sensitive to photo-degradation compared to the protonated format.

Studies reveal a 90% degradation rate in only 10 hrs of PCP in above-ground water, with pH levels of 7.3, whereas at pH 3 the assimilation rate was just 40% after 90 hrs. Once POPs are broken down into simpler forms by the soil, they become easier to transport across vast distances by runoff water and other carrier elements.

6.3.3 Children Living on Farms or in Agricultural Areas are exposed to

Pesticide drifts from sprayed fields: Pesticide drift is the accidental transmission of contaminants from their original source, it often results from off-target pollution caused by spray drift or runoff from crops or soil, which could cause human health damage, environmental contagion, and property destruction. Chemical spray drift refers to the movement of insecticide dust or droplets via the air during the period of application, or later on, to other venues apart from the intended location. Pesticide droplets may also be formed by spray nozzles, applied when spreading crops, forests, lawns or home gardens.

With localized spraying of a wide range of insecticides, wind drift can be reduced, plus considerable efforts are nowadays being made towards quantifying and managing spray drift coming from hydraulic spouts. Equally, wind drift is another way for transporting droplets of a correct size range, over to their destination through a vast area using ultra-low volume (ULV) spreading. There's a marked difference between exo-drift, the transmission of spray away from the target zone, and endo-drift, in which the active ingredient (AI) present in droplets reaches the target area, though it doesn't extend to the biological end-target. In most cases, endo-drift is significantly greater in capacity, and could, therefore, lead to increased ecological contamination, for instance, in cases where chemical insecticides contaminate underground water.

Figure 6.2: Bystander exposure is defined as a circumstance where people unintentionally get into contact with airborne chemicals.

Source: https://www.desmoinesregister.com/story/money/agriculture/2014/09/07/pesticide-drift-organic-farms/15230705/

Bystanders include laborers working at a designated area, different from where the chemical application is taking place, persons living in nearby zones of application space, or those simply passing by farms while they're being sprayed with pesticide.

Contaminated dust & soil: When obsolete pesticides are poorly stored, liquid pesticides may leak out from rusty drums into the ground. This groundwater may end up contaminating local lakes or rivers. Wind may further spread the pesticide powders across a vast area. Once pesticides are introduced into the soil, they spread fast at speeds that depend on soil type, the pesticide used, humidity content and organic ratio of the soil among other factors. A considerably small quantity of spilled pesticides may, therefore, cause a much larger proportion of contaminated soil. Case in point, only 30 tons of pesticide is enough to pollute more than 1500 tons of soil.

Soil contamination happens when dangerous pesticides are buried or leaked, or when they spread into uncontaminated soil. Pollution can happen during the improper clearance of harmful chemicals, or even when applying pesticides and fertilizers to crops. Moreover, these compounds may be spread if contaminated water travels through a region, usually carried by rain spill-off (Hallenbeck & Cunningham-Burns, 2012).s

If you suspect that your soil has been polluted, the first step is to create a barrier between the contaminated soil and other regions. Try to remove debris, spinning the soil over, sodding, casing with synthetic or cement, or digging and disposing of polluted soil to reduce exposure.

When working with polluted soil, try keeping the place moisture to curb dust accumulation. Also ensure the surrounding soil isn't contaminated. In particular, don't disrupt contaminated soil during windy periods, or when kids or expectant women are around since this can trigger adverse effects. Also, ensure the windows are closed during windy days, to prevent contaminated dust from entering.

Contaminated equipment and clothes: Personal protective gear is the final line of defense that protects a pesticide handler from chemical exposure. They consist of chaps, aprons, or even chemical-resistant suits. Sometimes, work clothes may also get contaminated with insecticides due to careless handling, packing, mixing or application of the product.

As such, you must handle and wash any clothing that may have been exposed to pesticides thoroughly, so that kids don't accidentally come into contact with it. Tossing contaminated apparels into the household washer or laundry bin, together with other garments is also restricted, since it can cause the transfer of chemicals to the other clothes and unconsciously to fellow family members.

Applicators can reduce the risk of exposure to families by taking certain precautions, some of which are:

- Spray garments should be swapped and washed each day.
- Before entering the home, remove any clothing soaked with lethal, pesticides residues, and throw away any sullied leather apparels as well. In some cases, the pesticide can't be entirely removed from the fabric, and full disposal is the only solution.
- Pre-rinse other apparels twice in a bucket of hot water. Discard the pre-rinse water into a septic tank, or use in subsequent pesticide applications.
- If the outfit can be laundered, inform the laundry person about the pesticide-contamination and reveal whether the chemical was in liquid or powdered-state. Including the suspected extent of the contamination.

Generally, modern pesticides are of less toxicity and results in less continuous damage in the environment, plus the quantity of active ingredient needed to manage pests is considerably reduced.

Treated fields: Nowadays, pesticides are widely used in sports fields, playing centers, and school grounds where they are applied to manage insects, weeds, and unwanted organisms. Roughly all publicly-run parks and children fields are spread with chemical pesticides, but they can cause side effects to children who frequently use these facilities. Typically, kids are exposed to symptoms through skin contact and breathing.

Figure 6.3: Children may be exposed to pesticides when playing in treated fields.

Source: https://www.thespruce.com/how-to-wash-pesticide-exposed-clothes-2147035

- Skin Contact: When children run in the field playing various sports like soccer, football or baseball, they're continuously touching the ground, either openly or through direct handling of the ball. Moreover, the herbicide drenched on the field can be absorbed directly via the skin, which is the largest organ in the body. Children are particularly vulnerable since they are much closer to the surface than adults. Equally, when they play, there's more sliding, dropping and rolling activities which put them at even more risk of contamination.
- Breathing: Kids can also get exposed to toxins through just breathing. When a youngster takes in air jammed with pesticide-laden dust particles, or vaporized herbicide, the inhaled air is absorbed into their bloodstream almost instantly, spreading to all body organs afterward. Pesticides that get attached to the soil, such as glyphosate, are particularly more prone to inhalation when children, having fun playing on the treated lawn to kick up the dust.

Moreover, exposure to chemicals found in playing fields, or the lawn can persist long after the kids have ceased playing at the field. Based on a 2004 study by the National Cancer Institute, it was found that recurrent exposure to some of these pesticides can increase the chances of developing cancer in the future (Matthews, 2008).

Parents spraying: Children can be affected by pesticides in what scientists call "indoor occupant exposure" phenomenal. Many homes use indoor broadcast-spreading pesticide canisters to control fleas and other insects. Studies reveal that some contain carbamates and organophosphates, which may cause moderate to severe toxicity in patients.

Most parents aren't aware that their kids may further be exposed to chemicals found in foods. When they spray insecticides in rooms or the kitchen but forget to cover any foods present then toxic compounds can get deposited on top.

Recently, in New York City, when there was a mosquito outbreak, many parents used a synthetic pesticide known as Resmethrin to try and control the situation. Also known as Malathion, this compound rapidly disintegrates in sunlight, particularly when exposed to the atmosphere and is mildly toxic to pets and humans. Following its application, many families sparked complaints saying that it triggered rashes and breathing problems, especially among the youngsters. Others also raised concerns that it elicited skin itching, headaches, runny nose, hoarse throat and dizziness.

Figure 6.4: Parents should always take precautions when spraying in front of their children, since pesticide poisoning has lots of short and long-standing nervous system side effects, such as irritation, sleeplessness, muscle weakness, breathing problems, anxiety, irritability, amnesia and depression.

Source: https://www.dailymail.co.uk/health/article-6832055/Mothers-exposure-pesticides-increase-risk-autism.html

Likewise, there are studies indicating that terminal low-exposure towards these chemicals could be hazardous, potentially causing cancer to children who may be exposed for extended periods of time. Still, many parents aren't cognizant of the potential toxicity of some of these pesticides to their children, meaning a lot of awareness needs to be done (Matthews, 2008).

Animal dips: Two of the most common contaminants present at animal dips are arsenic and organochlorine chemicals. Arsenic, in particular, is a carcinogen compound that's quite lethal to humans and livestock.

In most rural settings, plunge dips, shower dips and spray-booths are the most commonly used by farmers. These structures are usually adjacent to water sources or may be connected to a reticulated water-line for filling up the sump. If these points aren't properly covered, children may enter a contaminated area when playing and put themselves at risk of intoxication (Levine, 2007).

Similarly, leaks, spillage and splashes coming from the dip, poor discarding of compounds, and residues from treated animals may be left behind in places where they go, something that may result in cross-contamination if kids happen to visit the same areas in the farmyard.

Some common practices for eradicating spent liquids and residues from dips are;

- Gravity drainage to a lower surface
- Propelling or bucketing sludge from the sump
- Gathering of residual muck onto a "scooping mound"

Moreover, spent dip pesticides may be deposited into safe disposal units for later confiscation, instead of being left within the animal dip precincts where it can easily contaminate surrounding items.

Some of the spots likely to be polluted around an animal dip site are; the dipping shower, drainage platform and cattle holding yards. The land area most likely to be polluted is the site immediately surrounding the dip. As such, you should advise your kids not to visit these areas for their own safety.

Child labor: Generally, children are more susceptible to pesticide poisoning than adults, since they are still developing. According to ILO statistics, approximately 68 million kids worldwide are involved in child labor within the agricultural sector, with most of them coming from developing nations. Compared to adults, children working in these environments have diminished tolerance to harmful substances since they breathe, touch and interact with more pesticides in comparison to their weight. As such, their capacity to discharge harmful compounds is also lower compared to that of grown-ups. Exposure to dangerous compounds and pesticides can gravely affect their bodily and neurological development (WHO, 2010).

Likewise, youngsters have a lowered ability to evaluate risks. They might end up working or playing next to a pesticide-treated zone, just because they didn't read the warning sign. This issue becomes even more prevalent when considering that regrettably, children still make a considerably high number of rural agricultural labor force.

Figure 6.5: Children exposed to chemical poisoning when handing pesticide sprays.

Source: https://www.swissinfo.ch/eng/syngenta_swiss-made-pesticide-blamed-for-farmer-deaths-in-india/44408248

Side effects of pesticide contamination in kids:

- The youngster's brain, nervous makeup, and body organs are still undergoing development.
- Following contamination, a child's liver and kidneys won't be able to remove contaminants from their body, which means they may experience adverse poisoning that could potentially be lethal.
- Kids are also at a greater risk of pesticide poisoning than grownups since they take in more inhalations per minute plus have greater skin surface, in relation to their weight.
- Children spend lots of time near the ground, holding baseboards and grass where pesticides are regularly applied.

Due to this, it's crucial to reduce your youngster's exposure to pesticides, by applying an Integrated Pest Management (IPM) approach.

6.4 PESTICIDES IN DIFFERENT MEDIA

6.4.1 Air

Respirable particles contaminated with pesticides: According to research by Lelieveld, premature deaths caused by breathing of PM2.5 respirable pesticides in outdoor settings reached 3.3 million individuals globally in 2010, and this is expected to jump to 6.5 million fatalities by 2050. Air particles with an aerodynamic width lower than PM2.5 tend to be the most dangerous for human health since they are largely linked to breathing diseases as a result of their increased capacity to enter deeper into the lungs. Particulate matters from pesticides generally have atmospheric lifetimes varying from 3 days to one week. Within this period, particles can travel for 1000s of kilometers, based on the existing weather conditions, sometimes even reaching remote areas far away. This way, if insecticides or other dangerous substances get assimilated on the PM, then they could also be carried together by wind over vast distances, whereby populations located great distances far from the contamination sources would also be affected through inhalation (Philip & Landrigan, 2018).

Taking into account the extent of pesticides used globally, which is believed to be roughly 2 million tons of active insecticide ingredients, along their persistence within the ecosystem and bio-accumulative properties, this could be a major concern concerning air quality, as well as human health globally.

When applying pesticides, a great portion of the used dosage doesn't reach the final object, instead of being relocated to other zones through spray drift. Based on their physicochemical features, including vapor pressure, some pesticides could easily enter the atmosphere and be dispersed through gaseous/particle form (Philip & Landrigan, 2018).

Respirable aerosols during spraying: Some modern applications of aerosol sprays include diffusion of pesticides, remedial treatment of respiratory conditions, and general combustion of objects. These particles are known to cause respiratory problems when they enter human habitations.

The aerosol diffusion of FP elements is considerably influenced by the particle dimensions, scope and formulation type. In overall, larger powder entrainment is seen with smaller aerosol particles. Higher emission quantities are derived from drug-alone blends of spray-dried FP elements, and lactose mixtures of micronized FP compounds. Also, small variations in the particle dimension on the 1 - 10-mum scope exerts a large influence on aerosol dispersal of jet-milled or spray-dried FP compounds, which applies to both drug-based and lactose-only pesticide formulations. Narrow size dispersals of jet-milled or spray-dried compounds are highly crystalline. Variances in particle shape may also be witnessed between spray-dried and jet-milled compounds. Additionally, greater fine particle fraction (FPF), as well as comparative fine particle fraction, may be obtained from particles of diameters between 2-and-5 μm. Likewise, higher mass average aerodynamic width is associated with the type in ingredients found in the pesticide. Spray-dried particles further generated greater aerosol diffusion compared to jet-milled compounds. Plus even in narrow size categories, aggregation still plays a major part in aerosol pesticide dispersion.

Apart from these factors, there are various ingredients found in a pesticide that can either fasten or slow down dispersal, methyl bromide, for example, is commonly found is respiratory aerosols and is known to spread very quickly through air. Basically, chemical and physical pesticide qualities will determine its speed of distribution (Philip & Landrigan, 2018).

Vapor from volatile residues of pesticides: Pesticides vary in volatility from typical fumigants, like gaseous methyl-bromide, to herbicides containing vapor pressures of under 10^{-8} mm. Nevertheless, it's just recently that scientists have revealed that volatilization, as well as vaporized transports are key in the dispersal of 'nonvolatile pesticides,' such as DDT1 and multiple organochlorine elements.

Pesticide volatilization may be described as the transportation of pesticide vapors across the air. Different people ranging farm laborers to passersby may be affected by pesticides through breathing of vapors right after it has been applied. Volatilization is slightly different from the pesticide movement caused by spray drift. Pesticide vapor exposure as a result of post-spreading volatilization occurs from 3 main sources which are; volatile agricultural pesticides, indoor pest-control pesticides and semi-volatile farm pesticides (Philip & Landrigan, 2018).

Various factors will determine the extent to which pesticides are disseminated into the air. For instance, different pesticides have varying vapor pressures with different conditions. Heat further plays a big role here. Generally, pesticides have reduced vapor pressures at lesser temperatures. But as the temperature increases so does the insecticide's vapor pressure. Similarly, other environmental contributors, like weather and moisture content can also determine a pesticide's volatility. Insecticides can also be attached to soil, broken down by water or light, and get dissolved in water. All these factors determine their probability to become vapor. The method in which a pesticide is sprayed also has a role. Case in point, a pesticide consisting of small droplets, such as an aerosol, often gets into the air faster and more readily compared to a spray with bigger droplets (Capri & Karpouzas, 2011).

6.4.2 Soil

Hand-to-mouth behavior: Youngsters are more likely than grownups to be contaminated by pesticides, due to home applications caused by hand- and items-to-mouth contacts within affected areas. Nevertheless, new studies have generally confirmed mouthing behavior in kids less than five years in age. Hand-to-mouth contact can be defined as all actions whereby hands or fingers touch the mouth, except for normal dietary needs. The degree of oral exposure caused due to hand-to-mouth exposure is determined by the occurrence frequency, children who do this more often during the day are more likely to be poisoned than others.

Additionally, hand-to-mouth behavior is more witnessed indoors compared to outdoors. Plus for both cases, hand-to-mouth frequencies in inter and intra-personal vary in degree from 60% to 30% likelihood, respectively. The variance ratio among different researches is much larger than the mean, revealing that different studies consisting of different methodologies may bear similar core principles (WHO, 2010).

Another significant source of contact for youngsters is chemical compounds found in the toys they carry, for which toddlers usually put in their mouths when playing. If these toys are used in places where parents have already sprayed pesticides, then there's high likelihood that the children may get exposed when handling them.

Research shows that there isn't any relationship between mouthing regularity and gender. Nevertheless, a clear correlation is existent between the mouthing rate and age. Children under 24 months exhibit the highest potential of mouthing conduct, with more than 81% of those sampled showing it. On the other hand, kids above 24 months in age showed the lowest incidences of mouthing behavior whereby only 42% of observations were made (WHO, 2010).

Figure 6.6: Child playing in a field that may be treated with pesticide.

Source: http://www.panna.org/blog/tips-tweets-how-avoid-summer-pesticides

Most children have crawling behaviors and are highly likely to play near the ground. These behaviors increase their exposure to pesticide pollutants found in soil, dust, or carpets, including toxins that develop low-lying deposits in the air present in certain pesticides. Scientists further suggest that the use of pesticides affects kids' health more severely than adults. Youngsters may be especially vulnerable since their immune structures are still developing, thus may provide minimal protection compared to adults' immune systems. Likewise, their bodies are less capable of cleansing and excreting pesticides relative to adults (WHO, 2010).

The likelihood of having blood cancers conditions, such as lymphoma and leukemia heightens significantly in youngsters exposed to insecticides indoors. Even more, a correlation between insecticide exposure and juvenile brain growths has been witnessed. Youngsters who get exposed to indoor pesticides through consuming them are highly likely to be found with at least two forms of blood cancers. Also noticed is a positive yet not statistically substantial association, between infant household pesticide exposure and decreased childhood immunity (WHO, 2010).

Some of the toxic chemicals crawling babies may be contaminated with include; arsenic, chlorine, lead and bromine, including other heavy metals found in pesticides.

To reduce risk of poisoning, always keep an eye on your toddler to ensure they don't crawl into the contaminated zones, where you have recently sprayed pesticide. Children's car seats may also be a point of contamination, they could contain not just pesticides from sprays but other hazardous chemicals like flame retardants as well. All these contaminants combined are dangerous to toddlers.

6.4.3 Dermal Contact

Dermal means of pesticide exposure is common among swimmers, or rural-based children bathing next to contaminated areas. These chemicals usually enter waterways, such as rivers, lakes and dams, through rainwater spill off coming from surrounding agricultural fields. If children happen to enter the same waterways, they risk having these pesticides penetrating through their skin and causing health problems.

The typical mechanisms associated with dermal absorption, points of chemical entry, and environmental or physiological aspects affecting this procedure are well documented. Usually, lipophilic elements and adsorption of pollutants to water sediments are the primary factors that contribute to poisoning. Moreover, prolonged skin exposure, particularly under conditions that could enhance dermal absorption may cause toxicologically high amounts of some water contaminants getting absorbed. It's further suggested that swimming should be done within safe areas, in order to reduce the chances of children getting exposed to these hazardous chemicals unknowingly (WHO, 2010).

Compared to oral intake that's more toxic, dermal exposure is milder and symptoms may be seen within significantly short periods of exposure, typically just a few hours or some days after exposure. Liquid pesticide formulations are normally more dangerous than solid state particles since it's harder for solids to go through the skin or the mucous lining. Some acute health effects a person may experience are skin or eye irritation, or even respiratory tract irritation. However, on some rare occasions one may experience neurological and psychological health problems, such as endocrine reaction, mutagenic conditions and even cancer. While acute effects are quite simple to notice soon after poisoning, the terminal effects of insecticides take much longer to manifest.

6.4.4 Water

Pesticides in drinking water: tap, well: Pesticides can potentially poison drinking water supplies. Mostly, they are spread to farmlands, lawns and gardens, though they may find their way into underground water reservoirs, or surface waterways that nourish drinking water supplies. Whether these chemicals pose a health hazard is dependent on how poisonous the pesticides are, including the amount of water in question, and level of exposure that occurs daily.

Private drinking water reservoirs: It's estimated that around 13 million homes depend on family wells for daily intake water in the U.S. The risk of contamination from wells is higher than tapped water since the latter moves through sealed pipes while wells are open to ground-based contaminants.

Sometimes, natural disasters like flooding, landslides, and earthquakes may impact the overall safety of your well water by permitting impurities to enter the well system from outside. If you think that your consumption water well could be contaminated following a flood or other natural hazards, get in touch with your local or municipal health or environmental agency. They will provide advice on examining and testing the well. Plus, where possible, get in touch with a contractor with expertise in repairing drinking water wells, so that they can check, test and fix your well (Levine, 2007).

Safety tips for water wells:

- Put a good cap or sterile seal on top to prevent unapproved use of the water well.
- Keep correct records of maintenance history, including disinfection or sediment extraction, which may require the application of chemicals inside the well.
- Hire a qualified well-driller when doing any new construction, amendments, or closure of the structure.
- Don't mix pesticides near the well.

6.4.5 Food

Crops routinely sprayed: fruits, vegetables, grains: Pesticides are extensively used in the production of food, where they control pests like insects, weeds, bacteria rodents, mildew and fungus. However, routinely spraying fruits, veggies and grains place them at risk of contagion. Kale, in particular, ranked the 3rd worst affected food, immediately behind strawberries & spinach which were the worst contaminated. From a recent study, around

92% of kale products tested were discovered to contain at least 2 pesticide residues. While a single leaf contained more than 18 different pesticides within it. Among the pesticides commonly present in this crop is Dacthal. It's a potential carcinogen found on almost 60% of kale products tested. Over the years, kale has grown to become a very popular food, especially among health-conscious users. An increasing number of foods now have kale among their main ingredients, thereby putting more people at risk of pesticide contamination both directly and indirectly (Levine, 2007).

However, the vegetable's rising popularity isn't fully to blame for the product's propensity for pesticide pollution. Some effects could also be as a result of increased harvesting of the plant.

Figure 6.7: Pesticide pollution in food crops is a common problem worldwide.

Source: https://www.everydayhealth.com/rheumatoid-arthritis/diet/insecticides-autoimmune-disease-what-women-should-know/

According to the EPA, it's crucial to note that simply because a pesticide remnant is spotted on food, this doesn't mean it's totally unsafe to consume. Some negligible ratios of pesticides could still remain on fruits, grains and veggies, which further decrease significantly as foods are harvested, transported, washed, and prepared. But the presence of minimal pesticide residue doesn't mean the compound is completely at unsafe amounts, the EPA provides guidelines on this.

Bioaccumulation in animals and products: fish, meat, eggs, dairy: Scientists have discovered that dieldrin, Aldrin, DDT and BHC are some of the most likely pesticides to affect freshwater fish. Bioaccumulation of these chemicals in fish varies from one tissue to another, other factors that play a role include overall exposure time and type of fish species. Studies show that maximum accumulation of toxins first occurs in the liver, and then the kidneys. For instance, in some species maximum accumulations stood at 87 μg & 85 μg/gm of the fish's liver weight. When humans consume fish found in pesticide -contaminated zones, they risk being secondary victims of pesticide pollution (WHO, 2010).

The extensive production of broiler chicken is marked by confinement of large flock in small spaces. This may cause bacteria and diseases to flourish thereby making farmers spray pesticides on their chicken regularly. Nevertheless, while the control and eradication of pathogens is good, the use of certain poultry pesticides like quaternary ammonium, sodium hypochlorite, glutaraldehyde and organic acids may have harmful effects on those who consume poultry products. When used for pest control in poultry, pesticides must be applied rationally, always taking into consideration the allowable limits. The most typical treatment products are pyrethroid pesticides, which are regularly used, together with or in absence of poultry litter. Poultry feed should also be inspected for contaminants so that they don't end up in the animals' system upon consumption, and later transmitted to humans who eat them. Plus, ensure their nests are also kept pesticide-free in order to prevent skin and feather contagion.

6.4.6 Food Residues

Food residue contamination refers to the presence of toxic compounds in consumable items, which may cause illnesses. The effect of chemical pollution on overall health depends on factors like overall duration the food has been exposed to contamination and level of toxicity of the poisons. Some common chemical residues found in foods are; mycotoxins, acrylamide, organic contaminants like dioxins and heavy metals. Pesticides can enter a food chain through multiple sources. Pollution may occur during processes like planting, harvesting and processing of food. Likewise, chemical pollutants can enter a food-chain through secondary handling and processing sources, such as storage, transportation or even during cooking. Nevertheless, chemical pollutants are only harmful to human health when at certain high levels. It's, therefore, necessary to control their amounts in food, so as to minimize exposure and the risk of illness.

Common sources of chemical pollution include:

- Unwashed fruits and veggies. The pesticides or fungicides commonly used on farm produce to help them become healthy from infections are risky when consumed; this is why you must properly wash them before consumption so as to remove any pesticide residues that may be present.
- Vermin control products. Products like rat poison and fly spray are quite dangerous when consumed. Therefore, always keep them away from places where children can easily access, as well as far from food items where cross-contamination may likely occur.

There are various signs of toxicity one should look out for, if they suspect that their child has been poisoned. These include things such as; breathing problems, vomiting, trembling and difficulty breathing.

On average, there are well over 1000 pesticides applied by farmers globally to ensure food doesn't get damaged by pests. Every pesticide has its own special properties and chemical properties. Other cheaper (off-patent) insecticides, like lindane and dichlorodiphenyltrichloroethane (DDT), have the ability to stay in soil or water for several years. When crops are sprayed with these substances, or grown in soils containing them, they produce foods that may be harmful to consume. The harmfulness of a pesticide is dependent on its function, including other determinants. For instance, insecticides are generally more poisonous to humans compared to herbicides. This same principle also applies according to the amount of chemicals one is exposed to, as well as the route of exposure such as swallowing, direct skin contact or inhalation.

Extreme effects from pesticide poisoning in food residues happens only above certain safe quotients of exposure. When individuals come into any contact with huge amounts of pesticide, this could result in acute poisoning or extensive health effects, including tumors and intense effects on reproduction.

Mouthing: As part of their development, some kids develop the habit of consuming non-nutritive materials like soil, charcoal and clay. This condition is known as mouthing or pica. Nevertheless, mouthing has various chemical hazards, since the soil or mud your youngster is eating could be contaminated with pesticides. As an immediate action for consumed poison, take away any potentially hazardous residues remaining in the toddler's mouth. If the poison is a known pesticide, carefully read the bottle's label and follow the mentioned safety directions for unintentional poisoning.

Pesticide poisoning signs closely resemble other conditions like alcohol intoxication, insulin reaction, seizure and strokes. A few symptoms to look out for include:

- Swelling or redness around the lips and mouth
- Strongly chemical breath odor
- Drowsiness, and
- Mental confusion

With a toddler, if you suspect contamination, be on the lookout for clues empty insecticide bottles, packages or sprays, skin burns or stains. As a parent, ensure that you safely store pesticide bottles in a way that prevents leakage, or any other form of exposure that puts your kid at risk. Children who are in the mouthing stage will put almost anything they can find in the mouth, including chemicals that aren't properly stored. Apart from toddlers, other children below 5yrs old may also exhibit mouthing tendencies, which could be a sign of autism. They will bite at shirt cuffs, containers, tools and toys, some of which may have pesticide traces harmful to their health. To prevent this behavior, you can provide alternative 'replacement objects' such as fruits or healthy snacks, which they can bite on when the need arises.

Scope of the issue: Pesticides are considered among the main causes of demise by self-poisoning, especially among those living in low- to middle-income nations. Since they are fundamentally toxic and widely spread in the ecosystem, the utilization of pesticides requires firm regulation and control by authorities to avoid any human contamination. To deal with the issue, most regulators ensure that food which is sold to public complies with established pesticide regulations, particularly with maximum residue restrictions. Consumers can further restrict their intake of dangerous pesticide deposits, by peeling or washing their fruit and veggies before consuming.

Guideline levels of pesticides in food (MRL): In the U.S., the EPA sets tolerances regarding the maximum proportion of a pesticide that cab remains on food, which forms part of the procedures used for regulating pesticides. Moreover, in some nations tolerances are known as maximum residue limits, or simply MRLs. EPA evaluates fruits, grains, and vegetables among other foods to ensure they don't poison humans or the environment at large. Chemicals that pass this assessment are given authorization or "registration", which allows for their sale and usage as per the EPA requirements (WHO, 2010).

Before consenting to the use of any pesticide for food crops, the organization also sets a tolerance, or average residue limit, which refers to the quantity of pesticide deposit allowed to remain on a sprayed food commodity. When setting tolerance levels, some of the factors authorities look at include:

- The pesticide's toxicity and its break-down components
- Quantity of pesticide applied, including how regularly it's done
- Amount of pesticide which remains in food, particularly at the time or preparation and marketing
- Any likely routes of exposure by which the pesticide- contaminated food products.

During the assessment, authorities take note of the issue that diets of toddlers and kids may be dissimilar to those of grownups, plus that they partake more food for their body size compared to adults. These differences are addressed by blending survey data on food intake by infants and kids, with information on pesticide compounds to evaluate their exposure from different types of foods. While youngsters are at a greater risk of contamination, expectant women also fall under the category of most vulnerable individuals.

Guidelines to limit the population exposure: acceptable daily intake (ADI): Acceptable daily intake or ADI is a term used to define the maximum pesticide amount, which can be consumed each day over a long period, without any perceivable health risk on the consumer. Humans can be exposed to different chemical compounds via oral intake such as food consumption, taking groundwater, or hand-to-mouth transfer. It's vital to determine the overall amount of chemicals that can be taken on daily basis, in order to protect your health. Moreover, for pesticide deposits and food toxins, ADI may further be referred to as Tolerable Daily Intake (TDI). For food additives, authorities calculate ADI levels first, before comparing it with predictable exposure levels to find out whether the risks are tolerable or not. The measurement unit of ADI for pesticide substances is mg/kg BW, meaning mg for each kg of body weight.

The ADI is typically denoted from the lowest non- observed adverse effect level, commonly abbreviated as NOAEL. This figure is derived by putting into place a certain safety or ambiguity factor, that's usually 100, onto the NOAEL derived from the highly sensitive test species. In general, the 100-fold surety factor is dependent on the requirement for taking into account the variances in species, toxicokinetics as well as toxicodynamics.

Over the previous decades, pesticides, including other agricultural compounds have become increasingly common in food production. The accessibility of these elements has allowed for significant improvements in food productivity to be attained. Despite these positive factors, the pesticides have been found to cause secondary contamination through ingestion of farm produce treated with pesticides (Matthews, 2008).

CHAPTER 8

Toxicokinetics Varies for Different Types of Pesticide

CONTENTS

7.1 INTRODUCTION

People get exposed to pesticides in different many ways. This chapter is going to discuss some of these ways and related health issues.

7.2 ROUTES OF ABSORPTION

7.2.1 Dermal

Dermal pesticide exposure is when it enters through the skin, which is the most typical route of chemical exposure. People may get exposed to pesticides through splashing or mist during mixing, filling or application of the pesticide. Dermal contact may also happen when you handle equipment, clothing, or surface which has pesticide deposits on top.

Accidental contact: Dermal contact of pesticides mostly affects farmers and other agricultural laborers. In developing nations, farm laborers use outdated methods for spreading, like hand-pressure sprayers which expose their skin to more chemicals compared to other refined techniques. Additionally, those living in these areas have less knowledge of occupational hygiene standards. Meaning they are less likely to wear protective gloves and overall, which may otherwise protect their skin from direct contact with the harsh pesticides. Apart from the work environment, pesticide contagion may also be possible when undertaking activities like preparation, application, and cleansing of pesticides and their related equipment. To mitigate dermal health risks, users can apply the following measures: (i) Avoid modifying spray nozzles, this affects droplet size dimensions and escalates the degree of dermal exposure; (ii) Use appropriate work attire consisting of thick fabrics, mainly covering the upper body sections; and (iii) Thoroughly cleanse the tank sprayer prior to any application activity.

Acute dermal toxicity and outcomes: Acute pesticide toxicity is simply the chemical's potential to result in injury from just one exposure, typically of short duration. The dangerous effects that emerge from this single exposure, no matter the entry route can be referred to as "acute effects." In clinical studies, the acute toxicity level is considered as the overall amount or concentration that a toxicant has, for exterminating up to 50% of the test animals' population. The measure is usually written as lethal dose 50, or lethal concentration 50. Furthermore, the LD-50 and LC-50 values are dependent on just one dosage measurement, plus recorded in terms of mg/kg.

Occupational exposure: Farmworkers occupy the backbone of the agricultural economy. But this group of individuals is also among the least secured from job-related harms, such as pesticide exposure. Farmers are constantly exposed to pesticides in different ways. They may inhale or absorb toxic compounds via their skin during the mixing or application of pesticides, including other activities like planting, pruning weeding, thinning, irrigation, harvesting and crop processing. It's not just farmworkers who are in danger, but their families may also get exposed when staying in or around treated fields. Plus, studies further show that insecticides carried from the field to household on parents' apparel and skin, place their children at heightened risk of contamination. Though many cases go unreported, statistics on laborer poisonings show the magnitude of the immediate concern.

Reports have also shown increased risks of terminal diseases associated with pesticides, such as cancer, learning problems and birth defects. Similarly, undocumented farmworkers are not likely to seek health care when exposed to chemicals, plus they rarely report poisonings since this could put them under the radar of authorities. From the latest statistics, it is estimated that hundreds of farmworkers, including kids, encounter pesticide-related poisoning and diseases every year. Thoughtful reforms in the American immigration policies may be the first step towards addressing the appalling working conditions, experienced by 1000s of migrant agricultural workers each year. A healthy, secure and impartial system shall serve the basic safety needs of agricultural laborers, who mostly come from rural settings. Thereby, shifting away some of the harmful effects of pesticides.

Residues on surfaces: Many homes use household pesticides bought from supermarkets and hardware stores to control vermin, these are designed for both home and outdoor garden use. Though you can effortlessly obtain these chemicals from stores, it doesn't mean they're totally harmless. These could be toxic plus if improperly used, may affect a person's health, as well as their family, pets and surroundings at large. Residues of these chemicals may remain on top of surfaces, such that those handling them may get poisoned accidentally. Before buying any pesticide, ponder on whether your planned use of the substance is appropriate. Case in point, it might not be factual to expect your objects to be totally pest- free.

Hence, it's likely that the continuous use of insecticides could be more hazardous in the long run than even the vermin themselves. Even more, consider using non-chemical pest control techniques which are less noxious. Other useful tips are as follows: Ensure you identify any pest before buying a pesticide for it. Sometimes, what you presume to be mice in your roof might be another animal such as raccoons. If you're having difficulties pinpointing a pest, there are experts who offer vermin identification service. They will help you determine which creature has infested your home or garden so that you don't end up purchasing a wrong pesticide which only causes contamination; and when spraying, always use the slightest possible ratios of pesticide, since most of them are poisonous and may cause injury or illness when used incorrectly.

The probability of developing health problems relies on the kind of pesticide, including other chemicals which are found in the particular product you're applying, also the quantity you're exposed to, including total duration and how often you get exposed. The extensive use of artificial chemicals following the turn of the century has transformed agricultural practice. Insecticides are a wide group of assorted chemicals, commonly used to exterminate bugs, weed and fungi among other pests (Levine, 2007).

However, in spite of their introduction in the market, possibility of human toxicity has been high in many communities around the world. The degree of exposure is particularly great among individuals who are directly linked to the spraying process, such as farm workers. A few of these pesticides include hexythiazoks, azoksystrobin and imazalil, which were discovered on the skin and gloves of laborers. In general, the degree of active components found on users between sprayings was relative to the amount of active ingredients applied during sprayings. Moreover, there isn't any major difference in chemical concentration on the skin of workers, who engage in the tending and harvesting of pesticide-treated crops. Both activities present similar levels of dermal exposure to those who get exposed to the toxic substance (Levine, 2007).

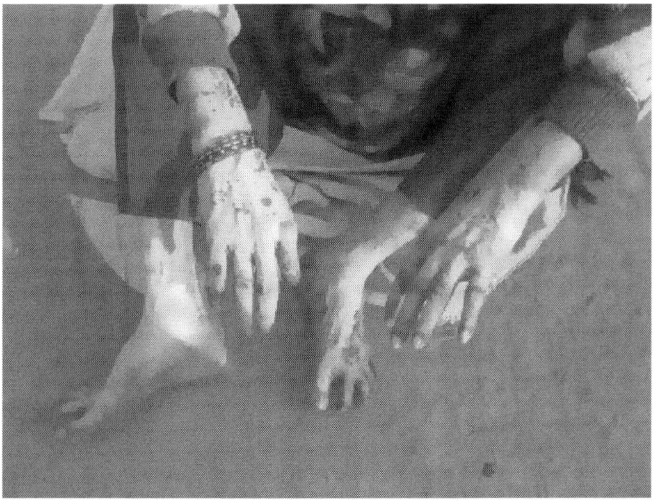

Figure 7.1: A farmer shows skin that has been affected by pesticide contamination.

Source: https://www.greenmylife.in/simple-ways-to-have-pesticides-free-food/skin-problems-caused-by-pesticide-exposure/

7.2.2 Ocular

Chemical contamination of unprotected eyes occurs from pesticide absorption in the ocular tissues. Ocular poisoning from pesticide exposure, as well as the dose-response relationship, is a fact that's been established among different societies. For agricultural workers, pathological side-effects of insecticides have been seen in cornea, lens, conjunctiva, optic nerve and the retina. Eye exposure is particularly prevalent in rural areas, whereas farmers who handle pesticides frequently put not just themselves in danger of ocular poisoning, but also their immediate family members.

Progressive lethal ocular effects resulting in defective vision are a major health concern. Farm workers are particularly at a greater risk of contact with pesticides and related ocular toxicity. Basic prevention tips can be applied to protect your eyes from pesticide-linked damage. It includes enhanced eye safety and precaution in workplaces, as well as effective pesticide regulatory measures by authorities for promoting public eye health (WHO, 2010).

The most commonly experienced outcomes of pesticide eye-infections are; itchiness, irritation, double vision, burning feeling and watery eyes. Blurred vision and burning feeling are the most common poisoning symptoms, though long-term ocular contact to pesticides can produce terminal eye effects as well. Beyond this, there are also studies to suggest that toxic effects could include visual neural pathways going from the brain.

Usually, it's the cornea, found in the middle of the eye, that's most affected by pesticide exposure. Over duration of 6 months, chemicals like Methyl Isocyanate which is found in most pesticides may already begin to damage the cornea (WHO, 2010).

7.2.3 Ingestion

While ingestion is not a common way of exposure, it can still result in serious poisonings. There are various reports of individuals accidentally consuming a pesticide that's been unlabeled or stored in beverage bottles. Laborers who handle pesticides can also accidentally ingest the compound when consuming food or smoking, especially if they didn't cleanse their hands first before touching the food. Pesticides and other related chemicals found in food particles and drinking water, pose an immediate risk to human health, these contaminants steadily increase in the environment, as well as inside the patient's body causing long-term diseases long after the initial contamination. However, there are disparities in defining terminal exposure and illness outcomes, considering the existence of vast sets of variables in terms of lifestyle, diet preferences, and occupation all of which contribute to the amount of exposure that a person experiences (Ohkawa, Miyagawa, & Lee, 2007).

In the U.S., about 75% of agricultural workers are exposed to pesticides. The problem is worsened in developing nations, whereby many old, non-patented, lethal, ecologically persistent, and cheap types of pesticides are used extensively, generating significant acute health complications, including local environmental pollution. Due to the nature of their work, farmers and farm laborers have a greater risk of contact with pesticides compared to other non-agricultural workers. Additionally, factors like the kind of pesticides employed, frequency of usage, and application method play a part in determining the extent of contamination. Additionally, the average duration of exposure will determine overall levels of toxicity that one has encountered. Furthermore, recent studies show that organometallic compounds found in some pesticides may have endocrine disruptive properties (Ohkawa, Miyagawa, & Lee, 2007).

Experts approximate that an average person is exposed to 200 artificial chemicals, including pesticides. For breastfeeding women, the situation is made worse due to their nurturing role. Meaning the toxins they are exposed to eventually reach their kids through milk. Among the most common pesticides mothers are exposed to include; dioxins and polychlorinated-biphenyls. Both of them don't assimilate into the ecosystem, they can persist within the environment for several years. Any tiny amounts of pesticides absorbed into the mother's body can accumulate over time, then later passed on to their kids through breast milk. The contaminants are usually fat soluble, meaning they get mixed into body fat and will build up, typically in the body fat over time. Your body normally draws on its existing fat reserves when producing milk.

In this manner, the minimal amounts of toxins present in your fat tissues may travel into the fat found in your breastmilk. But it's not only breastmilk that's contaminated. Pollutants may also get into other bodily tissues and liquids, including blood, and umbilical cord fluid, apart from breastmilk. Your child may only be affected by pesticide toxins if they are in high amounts inside your body. Otherwise, it would be considered safe if they continue taking your breastmilk. If the milk only contains negligible amounts of chemicals, then at that level it would likely be harmless to your youngster. To minimize the chances of baby poisoning, ensure that the food you consume hasn't been over-treated with pesticides, which could eventually enter your breastmilk (Ohkawa, Miyagawa, & Lee, 2007).

Children may consume a pesticide-laced substance accidentally when exploring their surroundings, or intentionally as a response to some underlying psychological problems. If this happens, you should contact a toxicologist immediately for direction on treatment. Diagnosis would be made depending on clinical evaluations made by the expert, including comprehensive laboratory studies to detect all swallowed pesticides and their possible risks. Nevertheless, as a short-term remedy, if kids present within an hour of ingestion any visible poisoning symptoms, activated charcoal may be given to temporarily manage the poisoning. The decision to give oral activated charcoal must be prudently considered and discussed with a physician so that it doesn't cause any further problems.

There are certain actions to take if your child consumes pesticides: The first step is to put your kid in a safe zone. Take him/her far from the pesticide- contaminated zone, so that they don't swallow more of it. If you own pets, try and secure them also in a separate spot. This way, they won't come into any contact with the pesticide chemical, or even your children during play. Also, don't try inducing any vomiting without an okay first from your doctors. Some ingested pesticide products are caustic, meaning they can trigger severe esophageal irritation when vomited back up by the child. In some instances, patients may require close observation for between 6 to 24 hours after consuming the poison; this is especially for high-risk cases where the toxin consumed is suspected to be lethal. Similarly, if the child was playing with toys in the poisoned zone, these toys must first be quarantined and cleaned off before being reintroduced to them.

7.2.4 Inhalation

Inhalation may happen when working around powders, air-borne droplet or vapors. The risk from low-pressure applications is considerably lower since most droplets are too big to persist in the air. Spreading an insecticide with high-pressure, low volume, or fogging farm tools can drastically increase the hazard since most droplets are small-sized plus they can be transported in the air for vast distances. Pesticides having a high inhalation danger must be marked with directions, showing that one should use a respirator when handling them.

Figure 7.2: Farmers' exposure through inhalation may also occur during pesticide preparation and application, including during the clean-up of spraying gear.

Source: https://www.dawn.com/news/1353014

Individuals who blend, load, and spray these compounds may be exposed to them through drift, whereby it's carried to their noses by air. Likewise, farmers may be exposed to chemicals when performing activities that aren't directly linked to pesticide use. Those who do manual labor in places treated with chemicals may face major exposure coming from direct spray, drifting from surrounding fields, or through contact with pesticide deposits on the plant or soil.

Inhalation exposure is often underrated since dermal routes of entry are normally the most common methods of contamination for farmers exposed to pesticides. Nasal exposure during pesticide usage occurs in situations where handlers leave their noses uncovered when handling the chemicals. By wearing a protective mask, you can significantly reduce your chances of being poisoned through inhalation. Likewise, farmers' exposure to insecticides can greatly be minimized through the use of lesser pesticides, including measuring the correct dosage every time before applying the pesticide (Capri & Karpouzas, 2011).

According to the latest studies, around 75 percent of American households utilize at least one pesticide brand indoors per given year. Another research suggests that approximately 80 percent of all pesticide exposure happens indoors. Among the most common contamination methods is indoor residual-spraying (IRS), which is a highly effective technique for malaria control, it involves the spreading of common insecticide on the inner walls of households to kill mosquitoes, thus interrupting malaria spread. Similarly, pesticides used for removing weeds in the garden may drift through air into your home, where they can be breathed by inhabitants.

In the U.S., pesticide use is controlled by the EPA which sets guidelines on the safety proportions to apply in each situation. For public places such as parks, they require that all personnel doing ground spraying should have proper training before being allowed to work. Including a competence certificate where possible. This will ensure that they apply the correct procedures to prevent spray drift, among other common methods of pollution. Moreover, should pesticide spraying be required in your home, a professional contractor having the appropriate pesticide credentials can be called. It's not recommended that untrained workers handle pesticides in public places. If the application of pesticides in treating a pest problem is unavoidable after all non-chemical techniques have been explored, it's vital that safety signs are put up around the venue of application to notify others of the spraying activity taking place. These are some of the effective measures to prevent inhalation and other forms of poisoning (Ohkawa, Miyagawa, & Lee, 2007).

7.2.5 Injection

Injection exposure usually occurs from suicidal poisoning through intake of organophosphate (OP) pesticides, they pose a serious health emergency and have high-mortality rates through inoculation. Most patients develop necrosis at the point of injection, and inflammation of the affected body part following injection. Similarly, a fasciotomy and clinical debridement may result in the patient but disappear within just a few days.

As society advances, the population encounters increasing societal pressures as well as family conflicts. These factors have contributed to high suicide rates among those who can't cope with such stresses, especially among young individuals, and consumption of agrochemicals is among the most common techniques they use for their actions. Moreover, in developing states, household pesticides are easily bought from stores, contributing to an increased occurrence of suicides. These chemicals are categorized as acetylcholinesterase inhibitors, chlorinated hydrocarbons and botanical agents. Injection poisoning is quite rare, compared to other forms of pollution like inhalation, but it still does occur in some instances (Capri & Karpouzas, 2011).

After entering the human body, organophosphates found in pesticides inhibit cholinesterase enzyme function; this compound is found in the nervous system and works by breaking down acetylcholine, which is a neurotransmitter responsible for transporting signals between the nerves and muscles. Organophosphates, being among the most lethal compounds found in pesticides, were invented in the 40s in Germany. While they're effective and don't pose any major environmental hazard, they also tend to be toxic and frequently used by suicidal injectors. At high doses, fatality following exposure can occur within just a few hours.

7.2.6 Transplacental

Transplacental toxin transfer refers to the transmission of pesticide-waste from a mother to a fetus through the placenta. Though not common, this may also happen through contaminated clothing that a mother wears. At the hospital, there are lots of toxins and chemicals plus the mother is usually given ward clothing to cover themselves. These attires are usually loaded with pesticides and chemicals from where they are kept or even poor hygiene by the ward hospital staff. When a pregnant woman puts them on, these chemicals may penetrate their skin and affect the developing child inside. They could also pass through the umbilical cord which is connected to the mother's circulatory system. Therefore, it's strongly suggested that multiple washing must be done on clothing polluted with concentrated pesticides, in order to remove excess residues (WHO, 2010).

Ward clothing offers protection to pregnant women against contaminants like pesticides found in the hospital environment. Ultimately, they will get contaminated with chemicals coming from daily use. Hence, you must carefully handle and wash the clothing to avoid any poisoning to the child, and yourself as the mother. Below are some quick points to consider when doing laundry for an identified pesticide:

Check the pesticide labels to find out if the apparels can be washed, plus how this should be done. If attires are heavily polluted, then laundering may not necessarily be a viable option. Discard any clothing or absorbent fabrics that are drenched or heavily polluted with the chemical's concentrate. Likewise, if the label permits you to launder attires that aren't badly contaminated, check for and read label statements giving direction on any special procedures that can be applied.

Acute pesticide poisoning can occur in children administered with topical application, used for treating head lice, scabies and other related conditions. Substances such as Diazinon-60, which has Class II toxicity levels according to WHO are among the most dangerous. In a recorded case, a patient treated with head-lice pesticides presented some neurological symptoms that necessitated quick medical intervention. Following emergency breathing and circulatory resuscitation, they underwent skin de-contamination and later dosed with atropine, high capacity oxygen and pralidoxime. Through scanning of electron micrographs on scalp hair specimens of patients, it showed both viable and nonproductive head lice nits which attach to the follicle shaft. The individual was hospitalized for a period of 7 days and cleared after complete recovery. This case showcases the significance of increasing awareness among health workers, including the community at large concerning the risk of misusing such pesticides during the treatment of insects.

Organophosphate pesticides are a class of insecticides that act on the body enzyme acetylcholinesterase. These are among the most widely used pesticides; they permanently inactivate this enzyme that's responsible for nerve working in insects, people, and many other creatures. Therefore, poisoned patients will experience difficulties in movement and control of vital body functions. Some common symptoms that patients may encounter are muscle exhaustion, lung weakness, hypertension and paralysis feelings. One may also experience cardiac arrhythmias, as well as heart blockage and QTc prolongation, these are sometimes observed in patients with organophosphorus agent poisoning. If you suspect poisoning from these substances, seek immediate treatment to prevent any further side effects (WHO, 2010).

7.2.7 Occupational Exposure

Occupational contamination of pesticides happens directly during the production of the food, transportation and storage, also in preparation and application by the user, during re-entry into spread fields, harvesting and farm equipment cleaning. In agriculture, majority of pesticides enter the victim's body dermally, closely followed by breathing and oral routes. Chemical inhalation mainly happens during fumigation, liquid preparation and input in closed environments, such as greenhouses and manufacturing stations. Work-related exposure of farmers, including other agricultural workers is probably the most significant contributor of pesticide pollution. Several studies have reported an association between work-based exposure to chemicals in the agricultural sector, and respiratory concerns. During the 1990s, it was reported among Tanzanian coffee plantation farmers that the rate of respiratory infection symptoms, shown between spraying and non-spraying methods of organophosphate pesticide application was quite high (Levine, 2007).

Similarly, another study was carried out in 2004 in Ethiopia, among farmers working in orchards, grain and vegetable crops within the Great Rift-Valley region. They consisted of pesticide sprayers, paired against a control group comprising of farm workers, and the pesticides used for treatment were organophosphate insecticide. The results revealed a significant declining in TSH, as well as drastic increase in TT3 and FT4 pesticide poisoning among the rural workers, in comparison to the test group. BChE amounts were also significantly lower for the exposed group, in contrast to the control set. These results show that farmers may be exposed to chemicals found in pesticides containing endocrine disrupting properties. Agriculture has been confirmed as the single most hazardous occupation when it comes to pesticide contamination (Levine, 2007).

7.3 BIOTRANSFORMATION

7.3.1 Into Inactive or More Active Metabolites

Most pesticides work by targeting structures or enzymes in pests that are identical or appear similar to those found in laboratory subjects. However, due to their toxicity, neither children nor adults are safe from pesticides that are improperly handled.

7.3.2 Active and Inactive Metabolites

A metabolite refers to a compound developed during the procedure of chemical reaction, through enzymes that are found in a cell or organism. The majority of pesticides are affected by metabolic degeneration, which occurs in plants or the ecosystem at large. Nevertheless, a portion of currently available chemicals, particularly insecticides and miticides, may be transformed into metabolites either within the pest or inside plant tissues.

There are occasions when pesticides may also have inactive or faintly active metabolites, these have less potency in terms of the enzyme reaction, therefore making them much more difficult to eliminate from the body as waste. Otherwise, pesticides can be metabolized through oxidation, hydrolysis, hydration, isomerization or condensation processes. No matter the process, the ultimate goal is always making it simpler to expel the pesticide from the body. Some of the enzymes present in metabolism of chemicals are already found in human tissues like the liver. Similarly, chemical metabolism rates may differ among patients (Matthews, 2008).

Generally, some factors that affect the reactivity of metabolites in pesticides are the type of chemical present in it and potency or concentration levels. In addition, most systemic pesticides are usually turned into metabolites which are water soluble, making them even more active in eliminating pests. Neonicotinoid systemic insecticides are a good example of these forms of pesticides.

7.4 DISTRIBUTION AND STORAGE

7.4.1 Fat Soluble Pesticides are Stored in Adipose Tissue

Adipose tissue is considered a storage compartment for various toxic lipophilic chemicals, such as pesticides. In fact, storing of pesticides in adipose tissues by the body is regarded as a protective activity, in that the chemical is kept in a harmless tissue, thus lowering its potency at the venue of toxic action.

Adipose tissue acts like a sponge by soaking up all toxic chemicals, which may otherwise be harmful when they enter the body. They are in the form of body fat that's easily being expelled from the body through sweating. For instance, chlorinated pesticides DDT, chlordane, aldrin and hexachlorobenzene isomers are lipophilic, meaning they are capable of accumulating in fat. On the contrary, during starvation, the fat that's been stored in adipose tissue may be assimilated and released into the system to provide energy; this could result in a drastic rise in blood chemical potency and pesticide availability for re-spreading into the body. Moreover, scientists have discovered that pesticides kept in adipose tissue could sometimes be swapped over by other chemicals. In lab studies, it was shown that DDT replaces dieldrin from the normal storage venues in rat adipose (Hallenbeck & Cunningham-Burns, 2012).

Chemical levels found in adipose tissue are a reflection of the intake history of a person. Basically, numerous factors have been depicted to affect fat tissue concentrations of pesticides, mostly being age and dieting habits. Also, the lipophilicity of chemicals depends on their elemental structure, whereby larger, more developed compounds are usually more lipophilic, as well as bio-degradation resistant.

7.5 ELIMINATION

7.5.1 Urinary Excretion

The urinary removal rates of pesticides like diethyl phosphate; including alterations in blood cholinesterase is determined by inhibition of serum or cholinesterase activities in red blood cells. Urinary excretion is the procedure of removing a chemical from the body through the renal system. Certain pesticides, which are biologically active by themselves, maybe removed in their original chemical form. Alternatively, other portions of the pesticide may undergo chemical alteration and get removed as organically active, or inactive, accelerating metabolites. While there are multiple routes for chemical elimination in the body, most drugs are excreted by pathways involving the kidneys. Renal emission plays a vital role in removing toxins or their metabolites through urine. A key characteristic of elements removed through urine is they're usually polarized or charged, as well as water-soluble (Capri & Karpouzas, 2011).

Chemicals that are lipid- soluble don't readily get extracted by the kidneys, plus they may undergo hepatic metabolism for increased water-solubility before possible urinary excretion. Likewise, chemicals going into hepatic-circulation state may further enter the bile, where they could get excreted from either the duodenum or small intestines. Also, based on a pesticide's chemical properties, it could get re-absorbed from small intestines before recirculation throughout the body in a process known as enterohepatic recycling. Apart from urinary excretion, in some instances pesticides may also be removed from the body via the lungs, sweat, saliva, tears, hair, or even skin. These are regarded as secondary methods for drug removal. Tests further show, there's no direct correlation linking the maximum concentration found in urinary insecticide metabolites to creatinine.

7.5.2 Biliary / Fecal Excretion

Biliary excretion is the active emission of pesticide molecules or metabolites stored in hepatocytes into an individual's bile. This is then moved to the gut, where these chemicals are excreted. The transportation method is alike to that of renal tubular secretion, even though efficiency levels of biliary secretion is rather variable. Even though many pesticides may enter the gut by this means, de-conjugating enzymes present in the gut, including the intestinal pH causes many chemicals to take non-polar lipophilic forms. These are then quickly reabsorbed through diffusion into the person's plasma. It's a process commonly known as enterohepatic re-use. Chemicals that undergo widespread enterohepatic cycling normally have extended durations of action (Hallenbeck & Cunningham-Burns, 2012).

Basically, pesticides are moved from the plasma system into the bile via hepatocytes. This transportation occurs due to accelerated emission of pesticides into the bile, through the help of organic anion transmission polypeptides (OATPs), P-glycoprotein transference network, and multidrug resistance-associated protein compounds (Mrps). Consequently, chemicals in bile pass into the gastrointestinal tract following storage inside the gall-bladder. This is then emitted from the body system through stool. Pesticides could be driven into the bile through a native method, or after metabolism this may be in form of polar conjugates. Likewise, the bile is emitted through the gut lumen, and from which native chemicals can be reabsorbed.

Also, in the guts, the enzymes present in the intestinal flora might hydrolyze the assimilated metabolites and release the active drug. After entering the intestinal mucosa, these active chemicals pass through the portal vein and then go back into the liver.

7.5.3 Excretion in Milk

The excretion of insecticides in women's breast milk is seen in regard to milk development, composition, feeding behavior and methods of pesticide transfer into milk. For a breast feeding woman, milk/plasma chemical concentration amounts are usually determined on the foundation of pH partitioning. Though some studies affirm these projections, others assume a need for considering extra factors like lipid solubility, or protein binding features of a chemical in milk to determine its excretion properties. Nevertheless, more research is required for specific pesticides, thus dosing via breast milk or the risk posed to infants remains an abstract subject. Many women worry about ecological pollutants found in their breastmilk, especially as the media periodically brings these concerns to the fore.

Any individual is at particularly high risk for unusual contamination to chemicals, including pesticides, herbicides and heavy metals. Additionally, in industrialized states, major spills or exposure may occur where mothers may be at risk of pesticide contagion in their milk. Studies conducted among Australian women show that, amounts of dioxin-like pesticides in breast milk have reduced over a decade by around 60%. A finding that's consistent with worldwide trends. Nevertheless, this milk may still have negligible amounts of dioxins due to its fat content. As such, nursing mothers are constantly warned about pesticide exposure, since many of the chemicals like dioxin can be found in breast milk. Even so, several studies done globally in places where chemical levels are high have revealed that breastfed kids are still healthier, compared to those that are given infant formula (Hallenbeck & Cunningham-Burns, 2012).

7.6 MECHANISMS OF ACUTE TOXICITY

7.6.1 Irritation (most of the pesticides)

Insecticide chemical irritation mostly occurs on the skin, where it causes cutaneous toxicity. When mixing, loading and applying pesticides, one's skin can easily get into contact with these products. There are many pesticides that can be passed through the skin, over into the circulatory system where they may cause toxic reactions. Depending on the amount one is exposed to, it could be enough to trigger severe reactions including itching and rashes. We stay in a world packed with insecticides and agrochemical substances, which are most irritating to the skin. We use most of these irritating compounds every day, but it's also good to learn how exposure can be restricted for your own personal safety. The dermis and mucous membranes that cover our body openings from the external environment, provide a protective barrier that retains water within the body, plus prevent external contaminants like bacteria, dirt, fungi and dust from entering the body. However, they can react to contaminants in the environment which drastically reduce their protective functions (WHO, 2010).

Primary Irritant Dermatitis: This form of dermatitis results from chemical substances that openly irritate the skin, such as caustic acids and bases present in some pesticides. Symptoms could redness and pain, severe blisters, or peeling and wounds forming on the affected parts. The spot of direct contact typically is the most intensely affected, plus it can be identified by checking for the presence of these particular symptoms. Treatment may comprise removing the irritant through thorough washing, topical ointments like steroid creams may also help relieve pain and scratching.

7.6.2 Allergic Sensitization (e.g. fungicides)

Allergic pesticide contact is associated with the environment and usually develops in stages. There's a period when a person could be continually interacting with allergenic compounds, but without showing any skin reaction. The period may last for just a few days or much longer. The allergenic reaction of a compound depends on its capacity to alter some properties on the patient's skin. This coating acts as a defensive cover against noxious substances. Some chemical compounds can remove fats, lipids and water from the skin. These elements diminish the protective role of the skin, thereby making it simpler for toxins to penetrate through.

Dermal allergy to pesticides starts with a procedure known as sensitization. After the diffusion of allergenic elements into the skin's outer layer, a procedure that could last anywhere between 4 days to 1 week. Allergenic substance proceeds to combine with the person's skin proteins. The mixture resulting from the allergenic compound and skin proteins might then be transported throughout the body through white-blood cells known as lymphocytes. Lymphocytes are a component of the immune structure, which protects the body against microbes or foreign substances. This system has the ability to distinguish and neutralize bacteria or substances detected several times.

When pesticide contamination patients are re-exposed, it's lymphocytes that recognize the particular microbe and interact with it. Plus, they also produce tissue-damaging compounds called lymphokines. The substances cause itching, pain, lesions, redness, and the development of small blisters on the dermis. Moreover, in severe instances, the pesticide allergy may spread to other body parts of the victim (WHO, 2010).

7.6.3 Enzyme Inhibition (e.g. cholinesterases and OPs & carbamates)

Enzyme inhibition is referred to as a decline in enzyme-related processes, production, and function. A set of clinically significant interactions happening between pesticide chemicals could result in CYP450 inhibition. These toxins are dissimilar in their selectivity for enzymes and are categorized based on their methods of action. Some chemicals are strong competitive inhibitors, meaning they contend for active sites against enzymes, despite not being substrates for the particular enzyme whereas other substances are non-competitive inhibitors.

Enzyme inhibition may trigger various adverse chemical processes that occur more rapidly, within a few days, compared to those experienced during enzyme induction. This is because they happen only after the inhibiting substance's concentration gets high enough to contend with the affected pesticide. In some situations, enzyme inhibition could cause potentially grave side effects; for instance, it could reduce the breakdown of terfenadine substrate, causing a protracted QT interval for the patient. Just like in enzyme induction, inhibition isn't limited to direct chemical interactions. Even widely assumed contaminants, like household insecticides inhaled during spray, may cause significant interactions within the body. This can result in increased plasma concentrations, particularly on the inhibited substance (Hallenbeck & Cunningham-Burns, 2012).

Some widely applied examples of pesticides like succinate dehydrogenase inhibitors, de-methylation inhibitors, phenylamides and quinone outside inhibitors are the major culprits of enzyme disruption. Also dangerous is cypermethrin, a compound, used to safeguard from various pests, such as moth pests of fruit, vegetable crops and cotton. These chemicals influence metabolism even at low doses by lessening or boosting enzyme activities.

Some have as well been discovered to be genotoxic, neurotoxic and hemotoxic.

7.6.4 Oxidative Damage (e.g. paraquat)

Pesticides can be categorized based on the kind of oxidative stress they generate; these chemicals affect the mechanism of free radical generation and the system's normal defense structures. They also have varying abilities to produce fat peroxidation or regulate body antioxidant status. It's generally assumed that free radical stimulation, induction of fat peroxidation, and disruption of the body's overall antioxidant functions are some of the side-effects of oxidative stress.

Oxidative stress can be defined as an acute imbalance of free-radicals, as well as antioxidants found in the body leading to cell and body tissue damage. While this process occurs naturally in humans, it can be exacerbated by pesticides like paraquat that are known to cause cell damage. Furthermore, a large set of research evidence shows that long-standing oxidative stress plays a role in the development of various chronic conditions, including cancer, heart problems and diabetes.

Naturally, the body's cells generate free-radicals in normal metabolic processes. Nevertheless, cells also create antioxidants that counterbalance these free radicals. As such, the body can maintain an optimal balance between anti-oxidants and charged radicals. Even so, various factors in the environment may overwhelm a person's ability to counteract oxidative poisoning, key among them being pesticides.

Sometimes, chemicals won't just act directly on cells, but also affect the natural immune reaction of patients. This kind of oxidative stress results in mild swelling, which may disappear after the immune mechanism combats infections or does cell repairs. Otherwise, uncontrolled oxidative stress triggered by pesticide exposure could hasten the aging process, causing a number of health problems to the individual (Hallenbeck & Cunningham-Burns, 2012).

7.6.5 Inhibition of Neurotransmission (e.g. organochlorines)

In order for the cholinergic nerve to obtain a passing message, acetylcholine should first be discharged from the receptor where it's been bound. This can only occur if the amount of acetylcholine present in the synapse cleft is minimal. Low synaptic ratios of acetylcholine may be achieved through a hydrolysis process, fast-tracked by the compound acetylcholinesterase. This element catalyzes acetylcholine to form choline and acetic acid. During acetylcholinesterase action inhibition, synaptic levels of acetylcholine will continue being higher than usual. If this hindrance is irreversible, which is possible with some lethal pesticides, the patient may experience sweating, paralysis, convulsions and bronchial compression among other complications (WHO, 2010).

The commonly acknowledged hypothesis suggests that organochlorines diminish cholinergic transmission within the patient's central nervous system. It further selectively and significantly impedes the acetylcholinesterase enzyme function, which is otherwise responsible for breaking-down acetylcholine.

The primary chemical reaction of organochlorines is believed to occur from this enzyme limitation. In addition, other likely targets for pesticide chemicals may include the impeding of various inflammatory communication pathways, which may exert further neurological damage. Generally, acetylcholine is the active neurotransmitter generated by neurons called cholinergic nerves. Inside the peripheral nervous structure, acetylcholine has a significant function in skeletal muscle action, including the moderation of body muscle and cardiovascular muscle. Also, within the CNS, acetylcholine is considered to be responsible for learning, memory, and emotion. Therefore, when this neurotransmitter enzyme is impeded through pesticide poisoning, the victim may experience cognitive difficulties in the mentioned areas. Other possible reactions are hypotension, severe nausea, bradycardia and seizures (WHO, 2010).

7.6.6 Uncoupling of Oxidative Phosphorylation (e.g. glyphosate)

Oxidative phosphorylation uncoupling is a mitochondrial level inhibition of the interaction between phosphorylation reactions and electron transmission. It results in the obstruction of ATP fusion, without affecting the respirational sequence and ATP synthase process. Miscellaneous components are often referred to as uncouplers, though weakly chemical uncouplers are of greater concern since they generally show quite potent activities. Currently, the strongest uncouplers that have been identified are known as hydrophobic salicylanilide S-13 and the limited phenol SF 6847 which are potent at concentrations levels of around 10 nM and above.

The potential danger of glyphosate is that it can impede transmembrane electrical processes in the mitochondria. Inhibition of uncoupled respiratory action is usually mediated through limited control of mitochondrial structures II and III, though not complex IV. The phosphorylative mechanism is affected through both a direct and secondary influence on the F0-F1 ATPase reaction. The adding of uncoupled ratios of pesticides to mitochondria has also been discovered to cause non-specified membrane permeabilization, that's shown by mitochondrial inflammation within the isosmotic sucrose channel. By and large, oxidative phosphorylation uncoupling is also connected to the weakening of mitochondrial cells, which are open to toxin infiltration due to the compromised surrounding film. Likewise, differences in toxicity witnessed may either be attributed to some by-products of oxidative phosphorylation, or a synergic consequence of glyphosate and derivative products. Considering that mitochondria are produced with a range of bioenergetic processes, essential for the control of intracellular aerobic power generation and electrolyte homeostasis. The findings show that pesticides can indeed be destructive at a molecular level (Levine, 2007).

7.7 ACUTE PESTICIDE-RELATED ILLNESS

7.7.1 Skin

Research from the U.S. shows that skin problems consist of around 35% of all job-related diseases. Out of all work-related dermatitis, irritant touch dermatitis consists of roughly 80 percent. Agricultural workers make the largest percentage of those affected by dermal irritation. While the specific body part which gets into contact with the chemical substance will determine how serious the skin damage is, also penetration of pesticides varies over different dermal regions. For instance, some chemicals pass through the facial area and upper-back regions much faster than the arms.

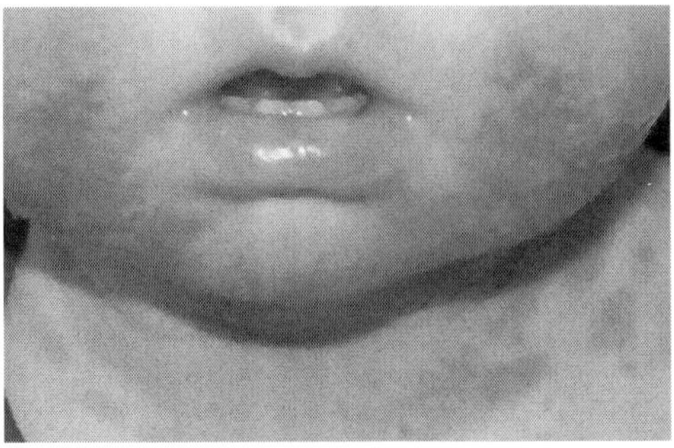

Figure 7.3: A child showing symptoms of grass rash or hives causes by playing on treated grass.

Source: https://www.healthline.com/health/grass-rash#the-takeaway

Factors that contribute to poisoning: The environment plays a crucial role in this matter. Case in point, hot and humid areas cause pesticide handlers to sweat. This perspiration can diffuse some kinds of chemical powders if they get into proximity with the skin. Moisture increases their noxious or irritant capabilities since solutions enter the skin much quicker than solid particles.

Work-related cuts or injuries on the skin also allow irritant chemicals to pass through more readily, thereby contributing to itching. Friction against the dermis may occur when handling farm equipment that chips away the skin, minimizing its protective function against irritants. Also, clothing soaked with pesticides is a possible contributor.

7.7.2 Eye

Pesticide burn happens when a toxic liquid or powder enters the eye. Notably, the injury occurs when a substance splashes across the face or rubbing your eyes following the handling of harmful chemicals. Based on the kind of chemical and extent of exposure, the potential for damage varies from acute redness and itching to full blindness. Always wear protective goggles when using pesticides to prevent such risks.

Figure 7.4: An Indian child loses her eyesight from pesticide exposure.

Source: https://theecologist.org/2012/jan/04/revealed-child-victims-pesticide-poisoning-india

7.7.3 Upper and Lower Respiratory Tract Irritation

Regular exposure of both upper and lower lung air-passages can cause significant respiratory tract disorders. Nevertheless, whether exposure to pesticides will translate into any significant respiratory disorder shall depend on its quantity and chemical composition. Though the number of possibly toxic, respirable chemicals is almost boundless, the pulmonic and systemic side-effects of various agents have already been documented by scientists.

Various factors are central in determining the site and extent of toxic lung damage arising from breathing of smoke, vapors or fumes with chemicals. These include the size and moisture solubility of the breathed compounds. The dimension of the inhaled chemical is a key factor of pathogenesis in pesticide toxin exposure. Larger particles get deposited in the nasopharynx from impaction and provide reduced infiltration to the lower lung chamber and alveoli. Alternatively, smaller chemicals pass through more readily and can reach up to the alveoli and bronchioles, generating respiratory poisoning at those sites. Compounds bigger than 80 μ in size are generally not diffusible through the nose, whereas those more than 5 μ won't usually pass to the point of the alveoli. On the other hand, very tiny particles like those less than 0.1 μ in width, enter deeply into the patient's distal airways, including the lung parenchyma (Hokkanen & Lynch, 2003).

The moisture solubility of the breathed material is yet another significant contributor to the dangers of inhaled toxins. Generally, substances with great water solubility, like sulfur dioxide, ammonia, and hydrochloric acid easily get deposited into the humid mucosal layers of the nasal, conjunctivae and oropharynx systems. Upon diffusion into the mucous membrane walls, these compounds develop acidic and alkaline properties which cause irritation.

7.7.4 Allergic Responses / Asthma (Fungicides)

In an American study, where dust samples from around 831 homes were inspected, it was found that residents in some of these homes developed asthma- related symptoms, due to exposure to insecticides and fungicides used to control harmful pests. Bio-pesticides containing endotoxins are typically used in the kitchen, living room and bedroom areas where inhabitants may get exposed and develop allergic reactions. The closest link between asthma exposure and pesticides was from chemical levels found in bedroom surfaces, and the bedding dust. Plus, this relationship was established both in adults and kids. When pesticides are used to eliminate mosquitoes, bedbugs and other creatures commonly found in bedrooms, they get deposited on the floors and other areas whey they accumulate over time.

Ultimately, continuous exposure to these pollutants increases asthma risk, plus it could result in lung inflammation as well. Exposure to crude pesticides causes abrupt bronchoconstriction in patients, as evidenced by many studies. However, this immediate bronchoconstriction can be hindered by taking certain preventive measures. Such as doing thorough cleanup of the home after every few days, so as to remove toxins that may be trapped on surfaces. Also, since diphenhydramine, being a histamine antagonist can reduce airway symptoms it shows that brevetoxins stimulate histamine-producing cells, including basophils and mast organisms (Matthews, 2008).

Moreover, the efficacy of atropine, which is an anticholinergic catalyst, shows that brevetoxins may activate neural pathways if used by a patient. The cholinergic pathway works by regulating the neurotransmitter acetylcholine, plus it's further stimulated by contact with organophosphate pesticides which are commonly used today.

7.7.5 Gastrointestinal Symptoms

Unintentional or suicidal consumption of pesticides can injure the esophagus and oropharynx systems. Some of them contain strong alkalis which severely damage the esophageal mucosa organ, plus there are cases where the stomach too can be affected. For instance, concentrated chemicals usually bypass the esophagus, but produce major scarring effects inside the stomach due to the transit time. The intensity of corrosive injuries on the stomach can be tied onto the specific agent that's been swallowed, and its overall concentration levels. Corrosive pesticide substances can be classified into different categories such as, fixatives, destructive elements, hydrochloric acids and softeners which have milder effects.

Side effects: Long-term exposure of chemicals may diminish the body's ability to generate stomach acid. Therefore, patients may eventually develop digestive problems, whereby the food they consume won't be able to be broken down into smaller particles for assimilation into the body. This condition is scientifically known as hypochlorhydria, patients may also experience nutritional shortages and gastrointestinal infections, even though quick treatment can inhibit any major complications. Furthermore, burns on the upper digestive tract, caused by corrosive chemicals, may result in severe stenoses, perforations, inanition or even demise of the patient. The kind of damage experienced eventually differs among different forms of pesticides. Upon admission after exposure, effort should be made to determine what chemical has been ingested. In some instances, the scope of the damage may be evaluated through an endoscopic method (Soloneški, 2014). Here, the doctor uses a specialized tool to observe the patient's digestive tract.

7.7.6 Neurological Symptoms

Neurologic damage is among the most widespread side effects of pesticide contamination. Chemicals like organochlorine, carbamate and organophosphates (OP) can directly infiltrate the nervous system, which is their mode of toxicity. Out of these pesticides, OPs are known to have the greatest reaction. Other varieties of pesticides, such as pyrethroid insecticides, fumigants, herbicides and fungicides have as well been linked with neurotoxicity. While acute poisoning syndromes have been well defined, there hasn't been enough discussion on the collective neurotoxic effects caused by low-level pesticide contagion. Apart from the neurological symptoms, terminal low-level pesticide contact has been associated with a heightened prevalence of conditions like headache, fatigue, dizziness, and shortness of breath. Additionally, OP pesticides, in particular, have been linked with multiple neurological abnormalities, such as risk of peripheral neural damage, lack of coordination, irregular deep tendon reflexes, as well as reduced energy levels (WHO, 2010).

Statistics show that mortality caused by neurodegenerative ailments may escalate by more than 100% by the year 2040. A recent analysis of the association between pesticide contamination and Parkinson's showed a positive outcome, even in cases where no dose-response interaction has been noted and no particular variety of pesticide identified. Other clinical tests have established delayed mental impairments that are relative to pesticide exposure, this may potentially be indicative of dementia, which generally starts early in individuals overexposed to pesticides. The cognitive disturbances endure in occupationally exposed individuals, even long after they have left the occupation that puts them at risk of chemical contact such as farming or livestock keeping (WHO, 2010).

7.7.7 Specific Syndromes

Cholinergic crisis (organophosphorus pesticides: Cholinergic crisis is explained as excessive stimulation of neurons at the neuromuscular intersection, due to increased amounts of acetylcholine (ACh). This causes inactivity or poor function of the AChE catalyst, which typically disintegrates acetylcholine. Among the most noxious pesticides are those containing organophosphate compounds, these are commonly found in commercial insecticides like phorate, ciodrin, ruelene, carbophenothion and dichlorvos. Mostly, the cholinergic crisis causes muscarinic signs including; increased salivation, urination, diaphoresis, lacrimation, gastrointestinal upset and hypotension. Patients may also suffer nicotinic effects, whereby skeletal muscle exhibits involuntary jerking and intense muscle contractions. It may further cause incapacity of cell membranes to repolarize, thereby causing fatigue and paralysis. Extreme reactions could further cause ventilatory failure or even cholinergic crisis fatality. Sometimes, exposure to organophosphorus pesticides (OP) can make muscles to stop reacting to the concentrations of ACh, this may lead to respiratory damage, flaccid paralysis, or other reactions like bronchial secretions and miosis or constricted pupils. If you notice any of these signs, contact your physician immediately for medical assistance.

Some properties of cholinergic crisis may be treated using antimuscarinic prescriptions, such as diphenhydramine or atropine. However, other conditions like respiratory arrest don't fall under this category. Atropine works by blocking muscarinic acetylcholine neuron receptors, but they won't improve the muscular strength or ability to respire in those with cholinergic crisis. For such patients, they may need neuromuscular blocking prescriptions, or mechanical ventilation assistance through endotracheal intubation until the crisis dissipates on its own. The respirational inhibition from cholinergic crisis currently only has invasive treatment methods, which some patients may find uncomfortable (WHO, 2010).

Bleeding (warfarin-based rodenticides): Warfarin rodenticide is a toxic over-the-counter, coumarin anticoagulant pesticide that's commonly used to exterminate mice and rats. It's classified among anticoagulants, which are natural and artificial compounds that prevent human blood from clotting. Warfarin-based poisoning occurs when an individual ingests the chemical accidentally. It happens when the substance is spread in or around a building, but without having any protective measures or warnings to those who may be around the area. Warfarin pesticide functions by inhibiting the working of Vitamin K1 in the body. This compound is needed for stimulation of clotting factors, which help stop bleeding in case of a cut. However, when warfarin deactivates Vitamin K1, the patient would no longer be able to develop clots, and bleeding may happen uncontrollably.

For victims, the most noticeable sign of contamination is hemorrhagic bleeding, which usually happens after 2-3 days following consumption. However, initial clinical signs affect the respiratory tract and include coughing or rapid and labored breathing. Chemical poisoning is tested by combining the history of its exposure or ingestion, visible symptoms, and conducting diagnostic tests such as clotting duration tests. These procedures are also useful for detecting anemia, and reduced platelet counts which are other possible symptoms associated with rodenticide pesticide exposure. Furthermore, the doctor may use radiographs or ultrasound to identify the presence of blood in the patient's chest or abdomen. Specialized assessment of the blood and liver may further confirm warfarin contamination, which is central in determining the Vitamin K1 treatment duration. Sometimes, if the chemical was just recently consumed, your doctor will induce vomiting and provide activated-charcoal to prevent additional absorption (Hallenbeck & Cunningham-Burns, 2012).

Caustic lesions and pulmonary fibrosis (herbicide, paraquat): Pulmonary fibrosis (PF) is an enduring and severe lung disorder. It causes lung tissue to become scratched, thick, and stiff thereby making it impossible for one to breathe. Even though there are prescription medications available, they can only minimize the rate of damage but not totally cure the disease. Compared to other body organs, it's only the lungs that are directly exposed to environmental factors. Every part of the pulmonary system, ranging from the trachea, bronchus, bronchi, bronchioles, and even alveoli are prone to damage coming from inhaled toxic pesticides. Among the breathing disorders caused by pesticide inhalation are reactive airway syndrome (asthma), bronchitis, chemical-induced pneumonia, and lung fibrosis.

Effects of inhaled pesticides on the lungs are dependent upon their physical and elemental properties, including host factors. Usually, large particles can't go through terminal bronchioles meaning they can't be breathed into the alveoli, or the air sacs responsible for gas exchange. Therefore, generally speaking, bigger, soluble components of toxic pesticides destroy the upper airways, whereas smaller, less solvable particles are more likely to injury the alveoli. Alveolar destruction is typically insidious, meaning it progresses from mere inflammation stage to the much severe pneumonitis, and permanent and lethal interstitial lung disease. Additionally, minerals and metals present in some pesticides are insoluble, meaning they remain inside the lungs and can cause injury directly over a long period. Conversely, pesticide gases are caustic in that they produce an inflammatory reaction in the lungs. These compounds can injure lungs in several ways such as chemical irritation, allergic response or soreness (Hallenbeck & Cunningham-Burns, 2012).

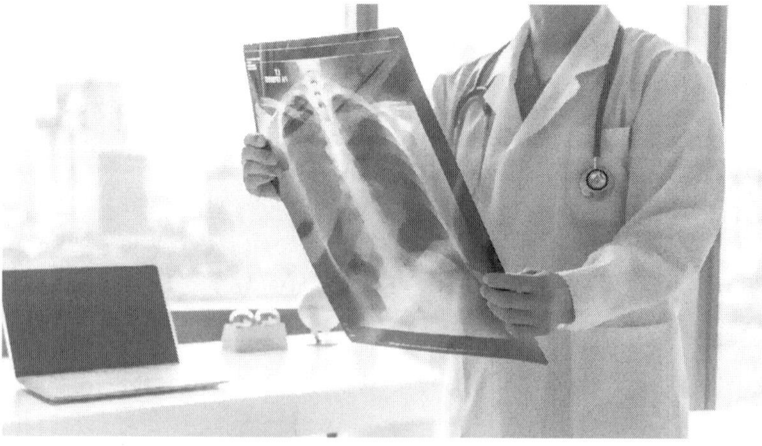

Figure 7.5: A doctor examines the x-ray image of a pulmonary fibrosis patient.

Source: https://www.dailytimes.live/health/understanding-pulmonary-fibrosis-04064088

7.7.8 Increased Risks before and during Pregnancy due to Exposure of Pesticide

Infertility: Scientists say that contact with pesticides can result in a wide array of reproductive illnesses affecting men, women and kids alike. In fact, a Harvard based study established that men who took food with greater pesticide deposits had reduced sperm count and lesser normal sperm. Similarly, another 2013 study associated legacy pesticides, such as organochlorine insecticide lindane, to a heightened danger of endometriosis among women. Women are less probable to reproduce when taking large portions of fruits or vegetables, considered to have heightened amounts of pesticide deposits. In fact, a female's odds of getting pregnant jumped by 79% and her chances of producing a live baby by 88%, once they swapped the intake of high-pesticide foods for those with less pesticide deposit.

Buying organic varieties of low-pesticide foodstuff such as oranges or avocados isn't the most ideal method to reduce contact with pesticides. A much reasonable approach would be reducing contact to high-pesticide veggies and fruits, such as strawberries or apples. The U.S. Agriculture Department has a wide list of pesticide exposed produce, which people should steer clear of. Some high-pesticide foods identified include; strawberries, spinach, grapes, tomatoes, kale and apples. Alternatively, avocados, beans, plums, onions, oranges and cauliflower carried the lowest amounts of pesticide residue. The scientists found that those ladies in their studies, who consumed the most pesticide-laced foods, were 18% less probable to get pregnant, as well as 26% less probable to deliver a live birth, as compared to women taking fruits and veggies with the lowest pesticide exposure. Prior research also suggests that pesticides can interrupt hormones responsible for pregnancy (Hallenbeck & Cunningham-Burns, 2012).

Perinatal death: Perinatal mortality is the demise of a fetus during the neonate stage of pregnancy. Exposure to certain pesticides spurs a high risk of this condition among women. During a study where fetuses got a moderate dose of pesticides, it observed just before parturition that 14% had passed away recently, while an extra 9% were seemingly expiring due to some severe cardiovascular symptoms triggered by toxin chemicals. Unlike infant mortality, where an infant passes away before they are 1yr old. Perinatal mortality usually occurs when the fetus is less than a month old. A fetus could be affected by dangerous environmental contaminants like pesticides which the mother is exposed to. These pollutants can enter their circulatory system via the placenta, potentially leading to harmful exposure that triggers perinatal death. Airborne pollution is among the most prevalent causes of fetal death. It usually comes in the form of particulate matter, which is spread through the atmosphere into the baby.

The most prevalent causes of death linked to airborne chemical exposure are fatalities resulting from respiratory infections, also known as Sudden Infant Death condition, or simply SID. Pesticides have further been connected to fetal death through spontaneous miscarriage, though more research is still required in this aspect. In a test where lab rats were given pesticides, it was found that 23% delivered stillborn pups. Moreover, out of those delivered alive, around 8% died after only 6hrs of birth, while another 13% perished within 48hrs. Upon birth, 14% of pups experienced first-degree coronary blockages, whereas 2% suffered from second-degree blocks. All young ones with second-degree blockages passed away shortly following their birth, whereas other fatalities were tied to cardiovascular problems (Levine, 2007).

Spontaneous abortion: Pregnant women, who stay near agricultural lands where pesticides get used often, face a greater risk of miscarriage or spontaneous abortion compared to unexposed women, based on studies done on California farmers. These studies are currently the most solid to date, showing that there are pesticides that can increase the threat of birth defects among humans. Unlike other substances used by agricultural industry, pesticides have been designed to exterminate living animals, which makes researchers especially concerned that they could harm humans.

Females who resided within a 2-mile radius of pesticide infested zones, where halogenated hydro-carbons were frequently used in the initial 2 months of their expectancy were at greater risk. These women were twice more likely to miscarry as a result of fetal defects. Similarly, other common categories of pesticides, like carbamates and organophosphates increase the risk of miscarriage by about 40%. Even though such epidemiological studies can't fully demonstrate that pesticides result to birth defects, they still raise a red flag. To further strengthen this case, the risk proved greater if females lived within a 1-mile radius of the agricultural land where the pesticides are being applied. These studies provide a step forward towards understanding the correlation between spontaneous abortion and pesticides. However, since most of the females sampled were in contact with multiple chemicals, it's not yet clear which compounds could be liable for causing certain symptoms. Part of the concern is that it's challenging to determine exactly the patient who was exposed, particular chemical used, and its amount. Mostly, researchers rely on individual's recollections of pesticide use and this may be insufficient to make a conclusive finding.

Premature birth: Preterm birth, also described as the delivery of a child before the 37-week full-term period, it's a common cause of fatality or long-term neurological dysfunction in kids. In America, the ratio of preterm deliveries has escalated by around a third since 2006. In fact, by 2012, it was projected that 1 out of every 8 kids was born prematurely, based on statistics from the Center for Disease Control (CDC). Among the leading causes of premature births is mothers' exposure to chemicals, particularly pesticides. Lab studies reveal a strong indication that chemicals, and pesticides, in particular, contribute greatly to these delivery patterns. There are many reasons to believe that pesticides could result in pre-term births; for instance, some of their chemicals may trigger inflammation in the uterus lining, thus making it difficult for the woman to carry a fetus fully. Other possible ideas regarding the mechanisms in which pesticides affect pregnancy may touch on females' immune response, hormonal disruption and oxidative stress. Scientists found that females who had the greatest amounts of pesticide metabolites within their urine experienced a greater risk delivering preterm. The chances were 2 to 5 times higher in comparison to women with the least levels of chemical exposure. Furthermore, when researchers assessed women who experienced "spontaneous preterm delivery," whereby a medical condition couldn't be identified for the early delivery, they established some correlation between exposure to pesticides and the risk of having preterm delivery. Residing or working within an area where crops are grown could expose you to larger quantities of pesticides, and during expectancy this may cause premature birth (Philip & Landrigan, 2018).

Fetal growth retardation: Fetal growth retardation is defined as a baby with less than 10% of normal fetal weight during their gestational age. This could result in considerable fetal morbidity or even fatality if not well diagnosed. This condition is typically caused by the insufficient maternal-fetal circulatory exchange, which results in fetal growth decline. However, with increased pesticide exposure in the modern age, chemicals in the environment may also contribute to this condition among pregnant women. Some chemicals are further known for causing congenital anomalies like trisomy 18 and trisomy 21. When fetal growth retardation is recognized, it's crucial to attempt and reverse any possible causes of poisoning. Fetal surveillance where delivery occurs before 38 weeks of gestation is also recommended. In most cases, infants having this condition will also develop cognitive and medical issues, even though long-term prognosis for challenged babies is also recommended. Studies show that pesticide exposure could be responsible for about 30 to 40% of fetal growth retardation cases, especially among females living around farms or working there. Continuous contamination of pesticides significantly increases the chances of acquiring the condition, as compared to women who only get exposed once or twice (Levine, 2007).

Before ultrasonography was developed, delayed fetal growth could be measured by lowered maternal weight increase, fundal height measurement and Leopold maneuvers. Presently, the condition is still measured on the foundation of fundal height dimensions. A significant delay in fundal height occurs when the pesticide causes a 4-cm or more difference, compared to what's normally seen in their gestational age. Moreover, according to WHO criteria, low birth weight can be recorded starting from 5 lbs and below.

Congenital malformations: Congenital anomalies are a major cause of infant and juvenile deaths, terminal disease and disability. It's estimated that around 2.68 million fatalities occur every year globally, during the neonatal stage. Pesticides are considered among the most prevalent causes of congenital malformations, particularly among women living in rural areas. Some of the most dangerous insecticide chemicals are as follows:

- Polychlorinated Biphenyls. This form of poisoning occurs from adulteration of foods by Polychlorinated Biphenyls, this results in symptoms such as dark-colored babies, gum pigmentation, CNS effects and abnormal skull calcification. Even though the threshold exposure level hasn't yet been identified, it's still likely to be teratogenic within small doses.
- Methyl Mercury. Exposure to this chemical may result in cerebral palsy, mental retardation, cerebellum hypoplasia and microcephaly. Contaminations have occurred from treatment of wheat and other crops, with mercury- laced pesticides used to prevent spoilage after harvest. Current environmental restrictions on mercury-use means it's unlikely to have a major teratogenic risk, however reducing the intake of crops grown with dangerous pesticides can be helpful.

Preconception screening is an effective way of discovering congenital malformations. It's particularly helpful in identifying babies at greater risk for certain disorders, or those mothers who risk transferring a congenital disease over to their kids due to heightened pesticide exposure levels. Screening could be done by obtaining familial histories, including carrier screening, plus it's particularly useful in nations where consanguineous marriage is prevalent. Ultrasound may further be applied to search for Down syndrome or other developmental abnormalities caused by chemical exposure.

Early childhood cancer: Around 20 years ago, scientists established that kids are by far more susceptible than adults to pesticide poisoning in the ecosystem. The finding sparked a belief that chemical contact early in life may cause cancer in kids, which may persist in their adulthood. In recent times, scientists have started to comprehend how certain pesticides trigger DNA alterations within the cells, thereby causing cancer in patients. DNA is the compound that constitutes our genes and controls nearly everything the cell does. However, environmental factors like toxic pesticides have been known to cause changes in this molecule, thereby initiating certain forms of childhood cancers.

Most childhood cancers aren't caused by family-inherited genes but instead develop from DNA changes that happen at some point in their life, or sometimes prior to birth. Each time a cell subdivides into two, it needs to replicate its DNA but this procedure isn't perfect and may result in errors, particularly when the cells are rapidly growing. This form of gene mutation may happen at any stage in growth, where it's commonly referred to as acquired mutation. Acquired DNA mutations in children begin from one cell where the pesticide has caused irreversible damages. This cell then transfers the mutation over to all the other cells reproduced from it. The transmitted DNA changes may stay in the child for long after exposure, sometimes even starting benign but appearing later on in life as full-blown cancer. As a parent, you should be careful not to put your child at any risk of pesticide exposure, which is an effective way of curbing childhood cancer (Levine, 2007).

7.7.9 Subtle and Permanent Effects during Brain Growth due to Pesticide Exposure

Brain structure and function: Researchers have identified approximately 12 developmental neurotoxins that can impact on a child's brain structure and function. A number of these chemicals can be found in everyday pesticides and include; ethanol, lead, chlorpyrifos, methylmercury, toluene, polychlorinated biphenyls and arsenic. Chlorpyrifos ,in particular, has grim side-effects on the fetus's brain development. Kids around the world are steadily being exposed to it, which has led to symptoms that are silently wearing out intelligence, truncating future achievements, disrupting behaviors and destroying societies. From 2006 up to date, it's estimated that the number of developmental neurotoxins affecting the brains of children has doubled due to advancements in technology and manufacturing. This finding is of great concern since it means more kids are likely to develop cognitive disorders, coming from exposure from chemical contaminants like pesticides (Philip & Landrigan, 2018).

Researchers have further noted that the numbers of diagnoses for autism spectrum syndrome and ADHD are escalating, plus that neurobehavioral development conditions presently affect about 10-to-15% of births. They further suggest that "sub-clinical decline in brain function," are becoming even more regular than typical neurobehavioral development disorders. Moreover, genetic factors contribute for only 30 to 40% of all brain-related development conditions, meaning the larger portion is due to non-genetic, environmental factors like pesticide exposure. In fact, there's strong evidence to show that these chemicals are responsible for a silent global pandemic of neurodevelopmental poisoning. Chlorpyrifos exposure, for example, results in significant deficiency in auditory function and fine-motor workings, but no shortcomings on more eyesight and inhibitory control functions. Organophosphate pesticides also appear to have the same effects on brain development.

Figure 7.6: A child eats strawberries that may be containing brain-damaging Chlorpyrifos pesticide.

Source: https://www.centerforfoodsafety.org/press-releases/4901/trump-epa-approves-continued-use-of-notorious-brain-damaging-pesticide

Neuronal and axonal differentiation: The nervous structure is a complex mechanism that affects sensory perception in humans. However, when it comes to contact with toxins some of these processes may be affected, and the person can develop nerve problems. Since neuron systems cover the entire body, some toxic outcomes could be specific whereas others generalized based on the particular part of the body where the toxins have reached. Neuronal and axonal differentiation after chemical exposure occurs in multiple ways, whereby the pesticide affects sensation, integration and response times. It will impede how the cell detects its surrounding environment, sorts and assesses information and finally the coordination of motor functions, secretory and neuroendocrine systems.

There are dangerous effects of pesticides: First, they interfere with chemical neurotransmission through synaptic dysfunction. There are 2 kinds of synapses, those found between axon and dendrite neurons, including those between the neuron and muscle cell. Pesticides can mix at certain points during the neurotransmission process, based on the way the chemicals react. This may result in either an escalation or decline in neurotransmission abilities. Chemicals can affect neurotransmission in various ways, for instance, some neurotoxins act by increasing or subsiding the rate of neurotransmitter chemicals within the presynaptic membrane. Additionally, some act by blocking receptors found at the post-synaptic membrane, thereby inactivating the neurotransmitter. Once a neurotransmitter passes through the synapse cleft, it normally binds with receptors present at the post-synaptic site. This binding process makes an ion gateway to open, altering the membrane potential for the post-synaptic nerve or muscular system or gland (Ohkawa, Miyagawa, & Lee, 2007).

Serotoninergic system: Serotonin is a central-nervous-system hormone that's responsible for regulating attention, behavioral patterns, and temperature. This compound is also produced in the intestinal enterochromaffin cells, where it controls processes like uterine contraction, gastrointestinal motility and bronchoconstriction.

However, when one is poisoned with pesticides they may develop serotonin syndrome, which is a possibly lethal condition related to heightened serotonergic action within the CNS. Many people who develop this disorder remain unaware of its causes and can trace it back to the pesticide they regularly use. Nonetheless, when left untreated it can have adverse effects on the serotoninergic system such as; cognitive system changes, autonomic reactions, and neuromuscular defects.

Patients with serotonin syndrome may experience a range of ailments such as; restlessness, anxiety, disorientation and agitated delirium. While questions still linger concerning the specific point where serotonergic symptoms related to pesticide exposure cause serotonin syndrome, the negative effects encountered largely depend on the chemical's molecular composition and toxicity.

When seeking treatment for serotoninergic system disruption, the doctor will likely require a detailed history of your previous chemical exposure, including the frequency of exposure and other related toxins that you might have been near to. Usually, cases of serotonin syndrome involving a monoamine oxidase enzyme are more intense and may result in adverse outcomes, including fatality.

Also, patients with accidental ingestions of chemicals usually develop higher toxicity and quicker serotonin damage action, compared to those exposed through other means like inhalation. In most cases, serotoninergic system disruption symptoms occur within 24 hrs after pesticide contamination.

Synaptogenesis: Synaptogenesis is a lengthy synapse formation developmental process, which involves synapse stabilization and activity-focused synapse refinement and removal, it's crucial for the creation of the neuronal system and accuracy of brain circuitry. Synapses are vital neuron components that allow data to pass systematically throughout the nervous structure, plus they help to regulate reaction to environmental factors and control body functions, memories, and sensory feelings. Any slight disturbance to the synapse system can cause brain illnesses.

In fact, evidence suggests that synapse dysfunction due to pesticide contact may be a key contributor to neurological syndromes. Some common neurodevelopmental disorders that patients may develop include; epilepsy, autism, startle disorder and Down syndrome, whereas neurodegenerative syndromes are Parkinson's and Alzheimer's (WHO, 2010).

Risk of epilepsy: Epilepsy is a synapse-related disorder characterized by an excessive coordinated electrical impulses of the central neurons. Apart from infection, traumatic brain impact and stroke, some scientists believe it could also be prompted by toxin pesticides. Usually, it's the temporal lobe that's most affected, though the hippocampal system and cortical areas are still considered prevalently exposed parts of the brain.

Some chemical toxins found in the brain alter the excitatory and inhibitory systems, which ultimately triggers the onset of episodic seizures.

Consequently, this triggers cellular and molecular alterations within the synapse system thereby causing epilepsy. It's also believed that pesticide chemicals can affect glutamatergic neurotransmission, thus leading to abnormal signaling and anomaly of excitatory neuronal transmission (WHO, 2010). Synaptogenesis is considered the primary indicator of optimal brain function since it involves improved synapse interactions and neurosensory pathways. However, synapses can be disrupted by pesticide chemicals which often lead to several neurological disorders.

Programming of synaptic function: The brain is a flexible organ whereby intrinsic CNS environments and extrinsic factors play a part in modeling and wiring neural networks. The development phase of a child marks a crucial period in central nervous system expansion, consisting of extensive improvements of neural networks, which are very sensitive to brain compromises like inflammation. Any exposure to pesticides can drastically affect the individual's immune system, including synaptic programming in the brain (WHO, 2010).

Emerging evidence shows that inflammatory cells can develop in the brain following pesticide contagion, and this has adverse effects on the synaptic structure and workings. Increased levels of toxins in the cerebrospinal system correspond to increased chances of having reduced brain-blood circulation, including minimized blood oxygen and this causes brain pathologies to trigger neuroinflammation.

In controlled lab studies, mice showed a heightened inflammatory brain response which lasted for around 3 days following insecticide contagion. Neurons play a crucial role in transmitting brain messages through chemical synapses. Any alterations in extracellular ion concentration caused by chemicals will disrupt neurotransmitter functioning and cellular behavior. Furthermore, toxins are known to inhibit synaptic vesicles from merging with the presynaptic layer of axon terminals (WHO, 2010).

Additionally, remote chemical exposure of a female during pregnancy may induce an early brain inflammatory response. In an experiment on fetal sheep, chemical injections into the amniotic sac during different stages of development showed brain development retardation. This is clear evidence that a youngster's synaptic system can be affected by toxins right from the prenatal stage. These chemicals also enter the hippocampus in babies where they could affect learning and memory.

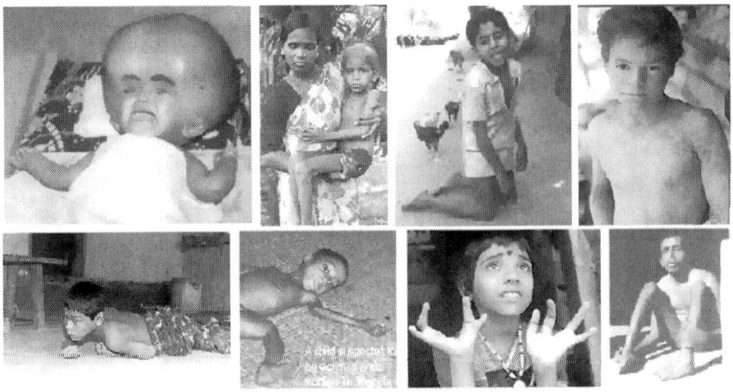

Figure 7.7: Indian children showing birth deformities caused by early exposure to pesticides.

Source: https://www.centerforhealthjournalism.org/fellowships/projects/epa-slow-halt-use-deadly-pesticide

CHAPTER 8

What Can Be Done to Prevent Pesticide Exposure and Poisoning

CONTENTS

8.1 INTRODUCTION

It is the responsibility of every pesticide user to prevent exposure at all cost. Pesticide exposure can be prevented at different levels. These include at local, community, national, and international levels. This chapter was prepared to offer vital information concerning these levels of pesticide exposure prevention.

8.2 PREVENTION AT LOCAL LEVEL

8.2.1 Use Pesticides only when the Benefits Outweigh the Risks

Avoid cosmetic or scheduled use of pesticides in the home

Pesticide applications should only be an option if other safer pest removal alternatives have been exhausted. If you must use them, ensure there is adequate ventilation in the room to prevent chemical congestion and reduce deposit levels. Sometimes, pesticide labels directly permit users to leave their windows open to prevent poisoning during spraying. However, if the pesticide is effectual only in closed spaces, don't stay there but instead apply the chemical and move out. Take all pets outdoors, including yourself and the family far from treated zones for roughly the duration of time suggested on the label.

Furthermore, avoid cosmetic applications of pesticide, but instead spray just on limited parts such as cracks where the vermin have been identified to stay. Avoid treating entire floors, walls and ceilings if the label doesn't expressly recommend you to do this. Excessive application of pesticides even in places that aren't affected by vermin can cause poisoning (Ohkawa, Miyagawa, & Lee, 2007).

Similarly, only mix the necessary portions you require for each spraying session. Don't prepare unnecessarily large volumes of pesticide to keep for future use, or so that you can apply higher concentrations for increased toxicity. When pesticides are stored for long without use, they eventually disintegrate and become dysfunctional over time. For rodents and insects, avoid using chemical baits that can easily be reached by kids or pets. But if you must apply them then ensure they are kept at an inaccessible point, where secondary contamination is least likely to occur. Finally, familiarize with the first aid tips mentioned in your pesticide bottle in case of any emergency (Ohkawa, Miyagawa, & Lee, 2007).

Use integrated pest management (IPM), non-chemical pest controls: Commonly known as IPM, integrated pest management is increasingly gaining popularity among farmers as an effective way of controlling pests. It's an ecosystem-based technique that emphasizes long-term pest prevention through a blending of techniques like habitat adjustments, ecological control, and spraying of resistant varieties. For biological management, a pest's natural enemies like pathogens, predators and parasites can be applied to the farm in order to reduce vermin populations. These microbes are generally safe to the crop but harmful to weeds and insects which could be present on your farm.

Cultural control methods are also useful in curbing pest establishment, dispersal, reproduction and survival rates. Case in point, altering irrigation practices may have an effect of changing habitat conditions, thus making it impossible for pests to reproduce. For instance, when water levels are altered in the ground then this can make certain harmful microbes and insects not to reproduce (Ohkawa, Miyagawa, & Lee, 2007).

In addition, mechanical and physical pest management measures can be applied to eradicate a pest directly, block them out, or cause the environment to be unsuitable for them. Rat traps are a great example of mechanical vermin management. Physical measures, on the contrary, include the application of weed management mulches, heated moisture soil-sterilization for disease management, or physical obstacles like screens to help keep birds and insects at bay. The most interesting thing about IPM is that it's applicable in almost every situation or environment, whether it's an urban, scrubland, agricultural or natural setting. It's chemical or pesticide footprint is also very negligible, which makes it safe for everyday pest-control applications.

8.2.2 If Pesticides are Necessary

Store in original containers with child-proof seals, out of reach, in a locked cabinet: To minimize poisoning chances, always keep pesticide products in their original packaging without transferring the components to another bottle. This way someone won't accidentally handle the hazardous substance, thinking it's a detergent or another similar solution. If possible, you can fix a safety latch on the drawer where the pesticides are stored, which shuts close after closing the door.

Some pesticides stored in single-use packets can be exceedingly concentrated and lethal, such that even a small quantity of the solution can result in serious respiratory, stomach ailments or eye irritability. Therefore, never allow your kids to contact or play with these packets. They usually dissolve quickly upon contact with water, saliva or wet hands (Ohkawa, Miyagawa, & Lee, 2007).

Similarly, don't stockpile any pesticide product. The best strategy is reducing storage needs, by purchasing only the appropriate amounts of insecticide which shall be required within a particular period. Stockpiling increases the chances of some pesticides expiring before they've even been used, this greatly escalates the chances of contamination since there would be lots of irrelevant toxic bottles around the home.

Furthermore, remember to store any flammable solutions far from your living quarters, or any ignition material like a car, furnace, outdoor grill, or even power lawnmower. Some pesticides have chemicals that are easily inflammable, such that even the slightest exposure to heat can set them ablaze. When this happens, home dwellers won't just be exposed to fire, but also the toxic fumes that are released during the combustion of chemicals. Also, don't store pesticides in places that are near food or medical supplies, due to the risk of cross-contamination (Ohkawa, Miyagawa, & Lee, 2007).

Educate on the safe use of pesticides: Before using pesticides, it's advisable for handlers to get some basic education regarding possible dangers and how to avert them. Some basic factors to consider include choosing the correct product and applying it based on the label instructions. Moreover, nowadays there are various programs which promote responsible and secure usage of insecticides that you can learn from.

They provide educational resources, teaching programs, and tools that are necessary for both private and industrial pesticide applicators to achieve accreditation and consequently to attain the knowledge necessary for safe and effective application of pesticides. The educational resources may further cover a wide array of pesticide safety subjects, such as pest identification, safe storage, ecological protection, pesticide drift, individual safety and clearance of pesticides among other important topics.

For everyday application it's advisable to stir pesticides outdoors or within well-ventilated rooms. Likewise, be ready for a pesticide spillage by having paper towels, garbage bags, non-absorbent gloves and other safety materials in place. These can quickly be utilized to manage the situation so that it doesn't get out of hand. Also avoid using too much water, since this shall only serve to disperse the pesticide even further, causing more harm to the ecosystem and those around the area. Moreover, when applying spray pesticides watch out for dusty or windy conditions that could spread toxins to other uncontaminated parts of the room. If possible, partially shut the doors or windows to your residence to prevent these elements from coming in. Another safety measure is to remove personal items, like toys, tools or clothing from the treated area to prevent chemical pollution (Ohkawa, Miyagawa, & Lee, 2007).

Follow the manufacturer's instructions: Most pesticide labels have clear information concerning how to apply the product properly and legally. These labels may further contain details on potential hazards tied to the pesticide, including instructions that must be followed in case of accidental poisoning or spillage. It's important to follow these label instructions since this shall reduce the risks and enhance the benefits.

As a user, always read the package carefully before buying any product and ensure it's applied for the specific purpose of manufacturing. Case in point, if a pesticide was meant for outdoor farm or garden use; don't try applying it inside your home assuming that it would have the same effects with household vermin. Instead, it would only put family members at a greater threat of toxins that shouldn't be in the interior.

Similarly, don't assume that a pesticide bought for one kind of treatment may be substituted for another different case. A fungicide will not be useful in eradicating cockroaches or ants and vice versa. Applying the wrong treatment shall only make the home environment full of dangerous chemicals, which don't even serve a purpose (Ohkawa, Miyagawa, & Lee, 2007).

Also, never take out a pesticide sticker from the bottle, or use unmarked pesticides since you don't know what chemicals they could be containing. During application, avoid using the chemicals in any way apart from what's expressively mentioned on the sticker. Always remember that the label or sticker is the most vital information source for pesticides, plus you must read and familiarize yourself with the instructions or possible warnings before applying the product.

Use protective equipment: Wearing a Personal Protective Equipment, also known as 'PPE', will greatly minimize the chances of dermal, respiratory, ocular and oral exposure, thus significantly minimizing the probability of pesticide contamination. These chemicals may have intense reactions on humans, with severity levels depending on the product's chemical composition and makeup, the amount that goes into the body, and duration of contact. PPE includes various items like safety suits, footwear, eyewear, gloves, aprons, headgear and respirators (Ohkawa, Miyagawa, & Lee, 2007).

Different pesticide brands require different private protective equipment. Note that any product containing a pesticide, such as baits, fertilizers and aerosols should be handled with the necessary PPE, and in a correct manner. Additionally, there are various PPE principles and beliefs that should be understood, in order to safeguard the health and wellbeing of everybody involved in touching the pesticide. Some of the situations where wearing protective equipment is necessary to include the following:

- Handling toxic pesticides and their containers
- Decanting or blending pesticides
- Dealing with a spillage
- Adding pesticides
- Entering a freshly sprayed or treated zone

When using protective clothing, always take caution during removal and wearing of the same, since the outer surface could easily be polluted with pesticide residue left from previous use. Moreover, the clothing must be cleaned down and removed after pesticide application, before final storage in a sealed, convenient, clean and dry bottle. Likewise, some items like the respirator are sensitive and need to be kept in their separate, sealed containers away from other protective equipment a person could be using (Ohkawa, Miyagawa, & Lee, 2007).

Respect re-entry times: Re-entry interval refers to the average duration of time between when the pesticide was spread to a region or crop, and when people can safely move into the area without wearing any protective clothing or equipment. The time measurement helps to safeguard people against contamination by pesticides, in case they enter a sprayed zone too quickly after treatment without having any proper gear on.

There are various ways an individual can get into contact with pesticides, but generally, it occurs by inhalation, dermal contact, food ingestion and eye contact with residue. The sticker on most pesticide containers provides details concerning re-entry intervals (Ohkawa, Miyagawa, & Lee, 2007).

However, sometimes a pesticide bottle label may consist of multiple layers of information. In case all these details don't fit onto the bottle panels then extra details concerning re-entry time could be found in a different booklet. Additionally, provincial re-entry directives may be available in stickers, seals, tags and leaflets which are attached to the container. Otherwise, re-entry times can be categorized into the following groups:

- Slightly toxic pesticide. Here, the pesticide may feature a skull & crossbones marker in trilateral or diamond shapes. It means there's a waiting time of between 24 hrs to 48 hrs if the product is administered as a wet-table powder.
- Moderate contamination. This is usually labeled as a skull and crossbones sign in diamond or octagonal shape, it means the substance is more toxic than the above and the average re-entry time is no less than 48 hours.

Generally, intervals are determined by the toxicity of active compounds found in the pesticide, method of formulation and the nature of work that involves pesticide application.

Pregnant women should not apply pesticides: Insect sprays, repellents, and various other pesticides are commonly used in the home to get rid of troublesome bugs. However, new research coming from the U.S. Pregnancy Association has discovered that unborn kids are susceptible to the toxic effects of pesticides. This especially happens when the mother is exposed to chemicals during their first trimester when the neural tube and nerve systems are developing.

Effects on babies:

- When mothers are in touch with pesticides, this may cause miscarriages, low birth weight preterm births, learning difficulties in kids and birth defects. Women who live or work in crop fields are often exposed to huge amounts of chemical contaminants. When expectant, it's crucial for them to avoid getting exposed to toxins which could affect the child's health.
- Developing fetuses are especially vulnerable, considering that they undergo rapid and immense changes during their 9 months in the womb. Any exposure to pesticides or toxins during these stages of development may cause long-lasting harm. In fact, a 2012 research linked prenatal contamination to a group of pesticides commonly referred to as organophosphates.
- They resulted in reduced birth weights and preterm pregnancies. Likewise, even in minimal doses organophosphates were found to be potent neurotoxins, with the ability to destroy the child's nervous system and brain. Children exposed to organophosphates

in the womb also develop lower IQ levels compared to the rest, which leads to learning problems as they grow up.

To limit your exposure as a mother, ensure that the foods you consume have been certified for consumption and thoroughly washed to remove any pesticides that may be dangerous to your child.

Figure 8.1: Some fruits like strawberries which pregnant women consume may contain pesticides.

Source: https://thefern.org/2019/03/what-pesticide-monitoring-misses/

Use the least hazardous chemicals, least dangerous mode of application: To reduce toxicity, substitution may be a viable option whereby you replace hazardous substances with much safer options. It's an effective method of reversing pesticide risk or replacing dangerous with non-dangerous compounds. Some chemicals contain carcinogens and mutagens which can cause irreversible damage to health upon exposure. Such chemicals are often described as substances of very high concern due to their adverse effects of users, plus most are persistent and bio-accumulative meaning they can stay within the system for several years after exposure.

Criteria for picking a safe pesticide:

- Safety is the topmost priority when searching for a least-toxic pesticide, the poison level of a chemical is measured on the basis of LD50, where the greater the LD50 figure the less toxic a pesticide is and vice versa. Also of importance is the residual lifespan of the pesticide, which is the duration of time it remains toxic in the environment before disintegration.
- The target pest species is also crucial. It's particularly important to identify the creature that has infested your home before using any

chemical substances to exterminate it. By all means, avoid using broad-spectrum pesticides that can kill or injure other beneficial species alongside the pest. In case pesticide use is inevitable, try performing spot treatments to minimize the chances of having other non-target organisms or persons affected. Likewise, speeds of interaction have a significant role in pesticide toxicity. Shorter lived, quick-acting and acutely toxic substances like organophosphates are generally more toxic, compared to slow-acting chemicals like boric acid which may not have severe side-effects upon human contact.

8.3 PREVENTION AT COMMUNITY LEVEL

8.3.1 Integrated Pest Management (IPM)

Homes: The integrated pest management approach combines everyday approaches and scientific ideologies to combat vermin, instead of the use of toxins. By using knowledge concerning the pest's behaviors, reproduction cycle, requirements and dislikes, it's possible to make environmental adjustments that will be intolerable for the creatures to live in but without using chemicals. Usually, food leftovers and a comfortable interior environment may attract a diversity of pests into your residences, such as ants and rats. However, by using IPM approaches you can make your household a bit more unconducive for these creatures, thereby curbing their populations. Among the applicable methods is closing any possible entries and exits that vermin otherwise use to gain entry into your home. The first part is identifying those particular entrance points, which may be close to the windows, piping system, or doors where these creatures can easily squeeze through (Matthews, 2008).

Additionally, find and remove any possible sources of food that could be attracting the pests. Store food for humans and pets in tight containers, locking the garbage container with a break-proof lid, and repairing faucet leaks that often create damp conditions which some vermin prefer living in.

Also, vacuum carpets and mats frequently in order to remove food crumbs, insect eggs and bugs which might be invisible to the naked eyes. Furthermore, it's advisable to eliminate any clutter around your residence, which only acts as hiding spots for vermin. You may also use a hot dryer to eradicate bugs, eggs and larvae which could be present in your clothing or bedding (Matthews, 2008).

Schools: Even schools and childcare centers aren't safe from pests, which usually lurk in the playgrounds and classrooms. Nevertheless, there are various integrated pest control methods that can be applied for long-term vermin control. They reduce the kids' exposure to chemicals that may otherwise be harmful. A good IPM system for schools can involve approaches that seek to modify the institution's environment, improve sanitation, and eliminate pest harborage spots, along with applying the lowest impact chemicals as necessary. Controlling risks from vermin and pesticide exposure are top priorities in a school-based IPM program.

Checking for insects:

- Insect monitoring can be done to identify their behaviors, levels of infestation and location. It's a sampling method used to record vermin presence and multiplicity over time. Monitoring, hence, has two key functions which are; detecting pest manifestation and recording population increase over time so as to find a suitable extermination solution.
- Traps might be set in a pest-infested area of the school, in order to detect signs of vermin before the situation gets out of control. During monitoring fewer traps are typically needed. A few sticky traps may be placed in zones that previously showed infestation signs, or where insects are most prone to entering the school building
- While vermin can enter a school at any time, they usually attack on a seasonal basis depending on their reproduction calendar. Case in point, during rainy seasons termites usually emerge in huge numbers, right from their nests found in various spots around the playground. If not properly managed, pest problems may result in destruction to school property, unappealing landscapes and other health hazards as well.

Public buildings: Public buildings are just as exposed to pests as private homes. Bugs, rodents and other vermin can be a major headache for owners of commercial buildings. Some of them can cause public health hazards due to the spread of infections, whereas others can damage structures and personal items. Even so, there are various steps that can be taken to prevent them from sparking any more trouble. These measures are as follows:

- Empty trash frequently: Most pests like ants, flies, cockroaches and mice are driven to regions that have garbage, where they often forage for food. Certain items like garbage cans give them an easy source of food, which makes them keep going back for more. If left unchecked, this can cause a large pest concern that's difficult to deal with. To minimize the chances of driving pests into your property, ensure all garbage-cans are discarded regularly instead of having them to stay overnight. The trash from within these cans must be placed in outside bins or trash cans plus tightly sealed. It helps prevent vermin from getting inside and nesting around your property.

- Remove wet spots and humidity: Many types of vermin are attracted to areas that provide sufficient moisture or water for staying cool and drinking. Once these creatures find such a venue, they will build their nest nearby or return frequently over a long period. In some cases, wet zones or moisture coming from faucet leaks and other problems may soften the wood, thus making it simpler for carpenter ants or other destructive pests to infiltrate and construct their nests.

Health centers: Health facilities like hospitals, terminal care facilities and physical or psychological rehabilitation centers, face increased challenges of sanitation caused by pests. For instance, bedbugs are commonly found in resting beds, waiting room furniture and even laundry equipment. In fact, as the health center's size increases, so does the risk of contamination since there are more entry points for vermin to enter. Cockroaches, in particular, are known to spread around 33 different types of bacteria, 6 varieties of parasitic worms, including multiple other human pathogens. Likewise, rodents can breach through a building using even the tiniest cracks or openings in the entrances (Matthews, 2008).

Prevention tips:

- Sometimes, a poorly managed plumbing unit in the home can attract pests like cockroaches and flies, which seek moisture in such damp areas. Fixing clogs, tightening floor drains and sealing entry points around the drainages can stop flies, cockroaches and rats from making use of the plumbing structure to move into the facility.
- Another possible access point for vermin is through food deliveries, where they may be carried inside food containers being delivered to the health facility by outsiders. Therefore, setting up protocols controlling food flow within the center is a crucial aspect of an effective pest management program. Case in

point, food service workers should examine all food shipments for real pests or indicators of pests, like droppings or destroyed packaging and food. Furthermore, any cardboard boxes where food is transported should be confiscated immediately after arrival at the healthcare center. Finally, ensure the laundry is kept clean in order to prevent fleas, bedbugs and other similar insects from spreading across the facility.

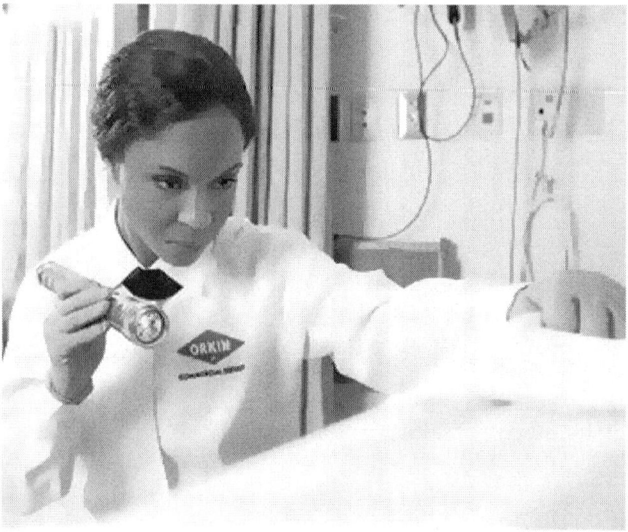

Figure 8.2: A health expert inspects for bedbugs hiding in a hospital bed.

Source: https://www.hfmmagazine.com/articles/1478-how-to-control-pests-in-health-care-facilities

Public parks: Nowadays, many public parks are embracing organic land management strategies, to help manage pest populations without using chemicals. These parks are run through two basic principles, which are organic lawn management (OLM) and integrated pest managing (IPM). They both seek to control the park in a low-impact manner, which is in accordance with natural ecosystem procedures and minimize the application of artificial fertilizers and pesticides which have adverse health effects on people.

IPM is basically an approach for scaling back on artificial fertilizers but instead relies on regular inspection, assessment and reporting to define a threshold for vermin populations. If that border is crossed, artificial pesticides can be used to restrict the extent of the damage. This same principle applies to weeds. To inhibit unwanted plants from multiplying, park managers usually come up with thresholds that are particular to the venue. This is done based on an understanding of the local species and designed to rationalize labor and pesticide applications (WHO, 2010).

On the contrary, OLM programs don't tolerate the use of pesticides, synthetic fertilizers or herbicides. Apart from close assessment and understanding of the native ecosystem essential in IPM, the OLM approach further utilizes organic fertilizers as a means of eradicating pests and weeds.

These are nutrient-rich, compounds got from animal or vegetable matter, like manure, chicken dropping or peat, they are full of nitrogen, potassium and phosphorus but no artificial chemicals that could be dangerous to people. Other effective methods of eradicating pests and weeds in public parks are; steam spot spraying, or boiling vinegar and applying on the affected parts.

8.3.2 Community Activities

Community campaigns: There are many community campaigns that have come up over the years, championing for the safe use or eradication of pesticides. Among the popular issues, they tackle is spray drift, which occurs when farmers spray their lands but the chemicals are carried away by wind to other far off places. If there are schools, hospitals or other institutions nearby, those found within these areas risk being victims of secondary poisoning. Most community campaign programs against pesticides are open to everybody, regardless of their socioeconomic background. You only need to have a passion for an environment free of pesticide toxins. In the U.S., states such as California have active community campaigns where parents, teachers, healthcare experts and others have come together to pressure farmers against the uncontrolled use of dangerous pesticides (WHO, 2010).

Within the same state, statistics show that close to 500,000 kids attend school within 1/4 mile of agricultural fields, where health-damaging pesticides are regularly applied to fruits and vegetables. However, through focused activism by members of the community, some farmers have begun to switch over these pesticides for other safer alternatives like organic varieties.

Even if you can't join a grassroots coalition, it's still possible to champion the course by contributing your own funds to a pesticide awareness program. There are multiple local organizations across the country that are fighting against unsafe use of pesticides, simply identify a campaign movement that you like and chip in the number of resources that you can manage. However, it's still possible to launch your own community campaign against unsafe pesticides, if you feel there isn't sufficient sensitization within the local community (Fitzhugh, 2012).

Figure 8.3: Community activists protesting against the use of organophosphate pesticides.

Source: https://www.thecalifornian.com/story/news/2016/12/08/public-input-into-pesticide-rules-ends/95168766/

School activities: In some areas, school-going children may be highly exposed to toxins, such as when the learning center is found next to a farmyard. Nevertheless, there are certain school activities that administrators can adopt to make much safer from pesticides. These chemicals can be found in a wide range of places like old light fixtures, classroom supplies, building materials, floor surface and on, just about anywhere it has managed to reach through spray drift originating from surrounding farms. Still, there are certain measures a school can apply to make the environment less toxic to children and some of them include:

- Cleaning or removing pests from the classroom using approved green products. During regular cleanup, where students usually remove cobwebs, wasp nests and other vermin from their classrooms. They can be encouraged to use organic alternatives to pesticides, such as vinegar which is a powerful homemade solution.

- A fan can be placed in the classroom to dissipate pesticide fumes spreading from the surrounding fields. This is a great way to prevent toxic buildup within the classroom that can cause poisoning.
- Building schools in safe zones. Before a learning center is constructed, it's advisable for the administrators to look for a site that's far from agricultural fields and other possible sources of contamination.
- Maintaining a clean and safe cafeteria. The school's cafeteria may also be a source of pesticide pollution, where the young ones may be served with foods containing pesticides, or fruits that haven't been washed properly to remove chemicals from the surface after being plucked from the farm.

Local awards or contests: Awards and contests seek to recognize the individual contributions of those actively championing for safe pesticide use in the community. Usually, participants are nominated by stakeholders in the environmental safety industry, who then proceed to vote for the most outstanding human-friendly pesticide activist. Those who've shown dedication throughout the years for the course stand a greater chance of winning environmental conservation awards. Their work should have impacted their local community to the extent that many people were saved from potential chemical poisoning.

Between balancing work and life matters, it's easy to get complacent on the harmful effects that these compounds have on people, including the safety measures that everyone must apply at a personal level to ensure they aren't affected by these toxins. In some cases, prizes may be awarded to groups or organizations that are involved in pesticide activism, and not just a single person. These awards are usually held on an annual basis, where the most deserving candidates are given medals and sometimes even cash to support their course. The nomination process may, however, vary from one award event to another. For instance, some allow for participants to nominate themselves for possible recognition of their efforts. Others only accept nominations done by local community members, towards an individual or group that they feel has contributed the most in safe pesticide activities (FAO/WHO, 2019).

Moreover, nowadays it's possible to register for a local award online by submitting your application through the event's official website. The panel responsible for giving awards will check your submission and determine if it's eligible for competing with others.

Pesticide-free "zones": There are various alternatives to pesticides that are still effective in controlling vermin, other than toxic chemicals. For farms or gardens, you can introduce beneficial insects that are harmless to humans and crops, but great at eradicating identified pests. A garden or its surroundings comprise of many insects that are actually helpful to the owner since they attack other bugs that are destructive. Therefore, it is important to learn how these beneficial garden insects may be identified, and find out where they can be applied to control vermin. Some of these beneficial insects are ladybugs, praying mantises, bees, dragonflies, predacious mites, green lacewings and predacious wasps among others. They can be bought from approved insect stores within your area (Fitzhugh, 2012).

Additionally, to prevent the larvae from digging into the soil which surrounds your plants, try using "collars" comprising of stiff paper, tar paper or thick plastic. Slice a foot square portion then wrap it tightly around the plant's stem, before pressing into the soil about an inch or thereof deep. Use a card clip to attach the piece into place. This shall prevent cutworms and various other digging insects from going into the soil surrounding your plants. Another 'pesticide-free' strategy is applying trap plants. There are bugs, which when provided a choice, will go to chew on one kind of plant but not another. Case in point, maggots prefer having radishes over maize and tomato worms would rather eat on dill over tomatoes. Hence, certain plants may be strategically put up to lure harmful bugs away from crops you want to protect (Fitzhugh, 2012).

Support organic farming: Organic farming provides a viable alternative to haphazard use of pesticides, which often lead to an increase in cancer rates within industrial farming zones. Rivers may become contaminated and the soil infertile. Some farmers are driven to support all-organic options bases on these concerns since pesticide residue has been known to taint fish, veggies and cereals.

Non-organic food generation makes increased use of chemicals which can contaminate water, the ecosystem and find their path into the food chain. In 2017, tests conducted by the U.K. government found chemical residues in about 47 percent of local food, plus most of these consisted of several pesticides and not just one. Prior to that, between 2011-to-2015, it was discovered that 100 percent of oranges including 86 percent of pears sampled consisted of multiple pesticide deposits. The best method to lessen your exposure to chemicals found in food is going all-organic. Approved organic food, such as fruits, vegetables, dairy, processed food and meat products shall generally consist of lesser pesticides (Fitzhugh, 2012).

Many individuals don't realize it but roughly 300 pesticides, which are routinely applied in non-organic farming, can be found in chemically grown food, despite washing or cooking. Alternatively, organic farming standards don't allow the use of synthetic pesticides or herbicides like glyphosate which are harmful to human health. Glyphosate is among the most widely used weed-killer globally. In Britain alone, statistics show that its use has jumped by an appalling 400 percent over the past 20 years. The chemical is routinely found in bread, in fact appearing in about 60% of all the tested whole-meal bread samples.

Figure 8.4: An organic farm where the use of pesticides is prohibited.

Source: https://allianceforscience.cornell.edu/blog/2017/11/organic-farming-can-feed-the-world-until-you-read-the-small-print/

8.4 PREVENTION AT NATIONAL LEVEL

8.4.1 Education Campaigns Aimed at Pesticide Users, General Population and Children

Pesticides can be hazardous if wrongly applied or managed. Coaching in their correct application helps to avert mistakes that could be made while enabling handlers of pesticides to reduce the danger to themselves, family members, community, and environment during their course of work. Pesticides are key for managing pest populations, but inexperienced applicators, improper protection and tools are some sections of concern which escalate the potential for harmful personal and ecological impacts coming from pesticide contact (Damalas & Eleftherohorinos, 2011).

Most governments require pesticide users to be educated concerning safe application practices, local regulations and proper storage of pesticides among other factors. Through training, agricultural workers can stay competitive by using modern pest control techniques and learn environmental issues affecting the sector as well. For large scale pesticide applications, it's mandatory that you should be trained if mixing pesticides for usage, for instance, applying water to form a concentrate. Those who calibrate equipment that is used in spreading pesticides further require training.

Mixers for aerial applications: If you often use pesticides operating as a mixer or loader, where aerial spreading is frequent then you should get an approval certificate from relevant pesticide control authorities in your country. They will require that you know the best practices for avoiding spray drift when spraying, including other practical methods for avoiding contamination.

Livestock pesticides: Livestock is usually affected by ticks and fleas and you'll need training if putting the animals through a dip-bath or spray race. This also applies when using pesticides for chicken and other birds that you could be keeping.

8.4.2 Restrict availability or limit use

Pesticides are useful in killing various classes of organisms. In fact, most just target a specific class or species of bugs, nematodes or mites. It's therefore appropriate to limit use to only the particular vermin that has been detected, and not generalize by applying a single pesticide for different pests which only increases the potential for contamination.

Pesticides marketed for use within the household are relatively less toxic, and mixed in solutions that aren't excessively strong compared to their agricultural counterparts. They should be fairly safe around folks, especially when used as commanded. However, children are much smaller than grownups, which mean that even smaller pesticide doses can adversely affect a child's body. These can cause behavioral issues, problems concentrating and even cancer cases. Hence, you should limit their use to only when it's absolutely necessary. Other nonchemical alternatives may be more viable for pest removal (Damalas and Eleftherohorinos, 2011).

For livestock, you may not require many pesticides when only treating animals having internal parasite, a process called drenching, or during external parasite treatments that don't need dilution or water mixing.

Another technique recommended for reducing pesticide exposure is integrated pest management, where nonchemical methods are used to control vermin populations. Case in point, gardeners may set off lady-bugs or any other insect predators onto their vegetation. They help to consume parasites that you would have otherwise used chemical pesticides to deal with. Likewise, you may grow plants that naturally draw birds or other predatory creatures. Cases include plants that produce pods or huge leaves that provide protective shelter for natural predators against pests (Damalas & Eleftherohorinos, 2011).

8.4.3 Establish and Monitor Maximum Residue Limits

Pesticide contamination is usually measured in terms of PPM, whereby 1ppm refers to one out of a million parts of the chemical. The particles that pesticides leave behind after a place has been treated are known as "residues". The top or maximum residue level refers to the greatest amount of pesticide leftover after application and vice versa.

To prevent poisoning, it's advisable to monitor the ratio of pesticides you apply following the FDA's standards. The regulatory organization has a residue monitoring system used to monitor the number of pesticide residues found in local and imported foods, in order to ensure they don't surpass the tolerable limits.

This is done through a multi-residue technique that examines roughly 700 different pesticide compound residues, applying a unique analysis and chosen residue methods that can identify pesticide chemical deposits not picked by other strategies like the multi-residue method (Damalas & Eleftherohorinos, 2011).

Apart from the above, you may collect plant or food samples routinely from your farm or garden to measure the amount of chemicals they've been exposed to. Focused sampling is usually used to check up on a suspected problem case, or to obtain residue information on other foods not commonly covered during regular residue assessment.

Whereas multi-residue testing methods check for the presence of multiple chemical residues within a single go. Selective or focused residue testing practices target a lesser number of particular pesticide residues, which are suspected to be at the location. At low levels, the chemicals may not be harmful to people even after it's been detected, measured and confirmed. But beyond the tolerable limits there's definitely need for concern (Damalas & Eleftherohorinos, 2011).

8.4.4 Surveillance and Epidemiological Vigilance for an Acute and Chronic Related Illness

Pesticides can trigger harmful health effects when used over extended periods or following repeated or constant exposure even at low levels. Such minimal doses don't usually have immediate effects, though, over time, they could cause chronic illnesses. In fact, long term pesticide contact has been tied to diseases like Parkinson's, asthma, anxiety, depression, ADHD, leukemia and cancer.

The absorption rate of a pesticide will determine how harmful it would be to the individual. More often, volatile chemicals and fine powders easily get inhaled compared to dense products or coarse granulated compounds. Concentrated solutions are furthermore more toxic than diluted ones and handlers should be careful when working with them, especially when there are children or pets around. They may come in the form of solid baits that are colorful, appealing and sweet, but could be easily consumed by toddlers. For some cases, if concentration levels of the reactive compound are below 2%, as is commonly experienced in pesticides for home use, the harmfulness in case of exposure could be because of the solvent used during mixing, such as kerosene or paraffin, but not actually the active chemical (Damalas & Eleftherohorinos, 2011).

Generally, pesticides have varying distribution and persistence levels in the ecosystem, even when distributed in any way through the soil, air, or water. This must be looked into so as to understand how acute and terminal exposure may happen since they are the primary mediums of exposure. Additionally, a pesticide's persistence and its ability to cause terminal symptoms will depend on matters like its physical or chemical features, including the environment's characteristics.

8.4.5 Treatment Capacities

Emergency services: Most pesticide products are inherently unsafe and they can be dangerous if you're overexposed to them. Also, accidents or misuse of substances may occur thereby leading to illness or injury. During such cases, emergency services may be required to save the person from potential fatality. The symptoms of acute pesticide poisoning will develop during or following exposure, they could take a few minutes or hours based on the route of exposure. Contact through inhalation usually results in the quickest emergence of toxic symptoms, then oral method and finally dermal contamination. Most symptoms of poisoning are cholinergic, meaning that the impact on the nicotinic, muscarinic, and central nervous system (CNS) receptors present in the body. During such instances, emergency services would be needed within the shortest time possible to prevent any worst-case scenario.

Patients should check out for muscular tweaks and weakness, which is often witnessed and may require respiratory assistance, considering that respiratory arrest could happen suddenly. Moreover, some pesticides are known to cause wheezing, coughing and chest tightness which can affect one's ability to breathe thereby leading to respiratory depression, unconsciousness and seizures. For organophosphate pesticides, the main cause of fatality is usually respiratory failure (Damalas & Eleftherohorinos, 2011).

Toxicity:

The potential for a chemical to cause severe harm will depend on the dosage. Those exposed to higher amounts risk greater injury, plus factors like age and weight of the individual shall also determine how their body reacts. Sometimes, internal injuries could also occur where a toxic compound damages the liver, kidneys, lungs or other vital organs. Such patients will require emergency services provided by a health expert.

Poison control centers: A poisons control center is a health unit that specializes in assisting with, prevention, diagnosis and managing of pesticide poisoning. The makeup and function of these institutions vary across the globe, however, for the minimum threshold they serve mostly as information and regulation organs.

- Roles & functions: A poisons control center will answer questions from visitors concerning ways of avoiding chemical exposure, including products that are likely to cause contamination when applied by a user. They may also provide an assessment to determine whether exposure to a particular substance is hazardous, or details concerning the treatment that ought to be given for certain situations. Most poisons centers seek to promote an evidence-based, affordable control of poisoning and further ensure that redundant or ineffectual treatment methods are avoided. In some regions, they maintain elaborate databases, such as toxicological records, product formulations and inventories of contamination inquiries made to the institution. Furthermore, many poisons centers engage in pesticide outbreak preparedness activities and response actions at both the local and global levels.

Data obtained from these institutions may be applied in reducing the frequency of poisoning by detecting emerging toxicological threats, stimulating preventive actions by chemical manufacturers and determining the efficiency of these measures. The poisons center data may further contribute to enhancing knowledge concerning human-related side-effects of these chemicals. These centers also contribute to establishing structures needed by local jurisdictions, to implement nationwide or international policies on pesticide regulation. Particularly, these include abilities for surveillance, vigilance and public response for health-related events concerning chemical agents. They basically act as sentinels for pesticide usage, by giving the public information and warning of possible risks.

Education of health care providers: Education will help healthcare providers to become more efficient at diagnosis and handling of pesticide poisoning. These programs are designed for all types of professionals who oversee these cases, including nurses, clinicians, and medical instructors among others.

The training courses available in this package may include; theory and practical lessons about how to take care of pesticide poisoning among patients, first aid tips and other necessary lifesaving measures. The education covers various aspects in empowering healthcare providers, extending from simple handling of materials to measuring the extent of damage caused by these pesticides on a patient (Damalas & Eleftherohorinos, 2011).

The diagnosis of any chemical poisoning case can't always be done with certainty unless the healthcare expert has a sound clinical judgment which is learned through education and training. The complex nature of some conditions, such as chronic contamination, should never be overlooked. The clinician also needs to bear in mind that various organ systems get affected by toxin exposure in different ways, plus that contamination may vary from acute to toxic depending on the concentration levels. Likewise, latency duration coming from disease exposure will vary, going from immediate reaction to extended or delayed manifestation.

Broad knowledge and understanding of toxicology are important for diagnosing environmental and work-related contamination cases. The criteria used are more or less similar to those applied for analyzing other medical problems, they may include:

- History taking, involving onset and sequential exhibition of symptoms, including observation of palliative and causative factor of pesticide contamination.
- Physical tests to identify symptoms and
- Laboratory results.

8.5 PREVENTION AT THE INTERNATIONAL LEVEL

8.5.1 Persistent Organic Pollutants (POPs)- Stockholm Convention

Stockholm Convention is an international accord for protecting people and the ecosystem, against persistent organic pollutants or simply POPs. These are pesticides that remain active in the ecosystem for long durations after use, getting widely distributed environmentally, accumulating in the fat deposits of creatures and generally being toxic to people and wildlife.

POPs spread globally and may cause destruction wherever they move to. To prevent this, The Stockholm Convention seeks to eliminate or minimize the discharge of POPs. Signatory parties have established a technique for dealing with excess chemicals, which have been determined as excessively hazardous. Eventually, the Convention leads the way towards a future that's free of toxic POPs. It further seeks to reshape the global economy's dependence on toxic chemicals (Yan, 2017).

In ratifying the Convention, nations will take steps towards eliminating or minimizing the discharge of these chemicals into the environment. Around 152 countries have adopted the Convention so far, putting it into law.

Essentially, The Stockholm Convention is best understood for having 5 key pillars which are:

- Elimination of toxic POPs, beginning with the twelve worst ones
- Supporting a transition into safer alternatives
- Targeting extra POPs for action

- Clearing old stockpiles and tools containing POPs
- Working together towards a zero-POPs future

Figure 8.5: A board of pesticide contamination executives in Bangladesh.

Source: http://www.fao.org/gef/highlights/detail/en/c/1200677/

Policies for POPs control: Like with all environmental programs needing resolute action from various stakeholders in the community, great situational analysis and all-inclusive consultation is necessary for the ultimate success of this program. As part of the Stockholm Convention requirements, each member state is required to develop a National Implementation Plan or NIP, which defines the country's methods or processes of applying the Convention's stipulations. In most nations, the UN provides assistance for putting the convention's policies into practice. This usually begins with 'Enabling Activities,' which are done for preparation and structuring of the country's first NIP. The assistance has been offered by the UN to 41 states, providing NIPs that are tailored towards country-specific factors and dependent on the region's unique governorship, policy and supervisory structures (Reigart, 2009).

Whenever new compounds are added into the Convention, members must first assess their NIP and input new information regarding how they'll handle the newly-listed POPs. This procedure of revising and updating the NIPs may be difficult for members lacking sufficient resources and technical abilities. Up to date, around 48 nations have been supported by the UN to conduct NIP reviews and update projects. Before adoption, the NIP Update procedure undergoes detailed baseline reviews, and inventories made into sector-focused action plans including their validation.

Decrease of Unintentional POPs emissions: The Stockholm Convention, in partnership with the U.N, proactively supports the application of best available methods and most ideal environmental practices within different environmental source categories as important contributors. These are regarded as the topmost crucial measurements, for continued reduction of un-intentionally developed POPs (uPOPs) releases like dioxins. Stakeholders have developed a wide range of projects designed for reducing and, where possible, eradicating uPOPs emissions from certain priority source categories in agriculture and other industries. The convention also emphasizes the participation and involvement of private sector parties in pesticide management.

8.5.2 Hazardous Chemicals and Pesticides- Rotterdam Convention (PIC)

The Rotterdam Convention (PIC) Prior Informed Consent was formulated on 10/09/1998 in Rotterdam, Netherlands and ratified officially on 24 February 2004. It covers around 150 countries that have already passed it in their respective jurisdictions. It provides guidelines on shared responsibilities between member countries, touching on importation and exports of pesticides and other chemicals among states, where established safety measures should be followed to protect citizens' poisoning. Certain banned or prohibited chemicals are also not permitted for trade, plus there's a mechanism in place for member countries regarding the mechanism used for exchanging information concerning potentially hazardous compounds (Reigart, 2009).

Basic principles of the convention: The treaty provides a means for lawfully acquiring and disseminating information, for informed decisions to be rendered by importing states regarding whether they want to get future shipments of various identified chemicals, plus ensuring compliance by exporting countries concerning these particular decisions. Member states are expected to exchange information with each other, concerning potentially dangerous chemicals which they intent export or import from other jurisdictions. The primary goal of this convention is to offer technical help for developing nations, and countries with systems that are transitioning, to formulate the necessary infrastructure and capacity for implementing the requirements of the Convention (Reigart, 2009).

In total, there are 53 chemicals that have been mentioned in the Rotterdam Convention's prohibition list, 36 of which are pesticides and 15 industrial compounds. Among them are Chlorobenzilate, DDT, Aldrin, Binapacryl, 1,2-dibromoethane (EDB) and Endolsulfan just to mention a few. It's also expected that more pesticides would be added into this list, as their toxic properties continue being discovered by researchers.

(PIC) Prior Informed Consent Process: For compounds to be included in the banned category, a Prior Informed Consent (PIC) process should be followed. Upon addition, a special 'decision guide document' (DGD) having information on the chemical, including the regulatory plans to ban or intensely restrict the compound for health or ecological reasons, would be distributed to all Parties.

Members have around 9 months to provide a response regarding the impending import of the substance. The import reply may consist of an ultimate decision to permit for import of the compound, deny its import, or only allow it subject to certain preconditions or a provisional response. Additionally, the import verdicts are usually circulated by PIC circular following every 6 months, plus exporting nation Parties are required under the treaty to take active measures towards ensuring that exporters found within a particular jurisdiction conform to the decisions.

Information Exchange: Apart from establishing the PIC formula for blacklisting chemicals, the treaty promotes the mutual exchange of data on hazardous chemicals. It requires exporters of those toxic chemicals to utilize proper labeling, as well as guidelines on safe-handling, while also informing importers in case of any imposed restrictions or bans. Developing nations, in particular, are required to inform other parties when they're experiencing problems, triggered by severely toxic pesticide blends under use within their borders (Reigart, 2009).

Ever since its development, the Rotterdam Convention has played a big role in managing the production and supply of harmful pesticides throughout the globe.

8.5.3 International Organizations: WHO, IFCS, FAO, ILO

The World Health Organization (WHO), in partnership with the Food and Agriculture Organization (FAO), signed a treaty on cooperating in the effective control of pesticide usage. The two international institutions are united on providing reliable, coordinated and consistent counsel and support to member countries and other stakeholders regarding sound management policies.

These regulatory bodies have a code of conduct which shows the mutual responsibilities of various segments of society in implementing the policies, including the state, manufacturers, global institutions and so on. They provide a framework governing the usage of all pesticides, including those for agriculture and general public health (Copeland, 2007).

Objectives of the international organization codes: The key role of these regulatory codes is to set up voluntary models of conduct touching on both public and private institutions, which are involved in handling of pesticides in one way or another. They particularly fill in the gap where there's inadequate or non-existent legislation on regulating pesticides. Some of the bodies addressed by this statutory include; pesticide traders, global organizations vermin control operators and governments. The policies are designed for implementation within the framework of global legislation, coming from the basis whereby appropriate entities cited by the statutory may determine if their proposed actions or omissions of other factors consist of acceptable practices. Also, there are descriptions of the collective responsibilities of various segments of society to mutually work towards achieving certain pesticide-reduction benefits. Governments are further encouraged to adopt a cooperative effort of minimizing exposure, during exporting and importing of chemicals across countries. While also promoting practices that reduce potential health and ecological risks tied to pesticides, and ensuring safe and effective use among citizens (Copeland, 2007).

It's also recognized that appropriate training within all proper levels is necessary for implementing and discerning the regulatory code's provisions. Hence, entities mentioned should give top priority to relevant training and capacity development activities related to pesticide management.

Member states are required to encourage practices that minimize risks throughout the pesticides' lifecycle, with a goal of reducing adverse side-effects on people, animals and the ecosystem. This is inclusive of the prevention of accidental poisoning due to improper handling, storing, movement, usage or disposal, plus from the existence of pesticide particles found in food and feed. Furthermore, states must ensure that the chemicals their citizens use on a regular basis contribute towards sustainable enhancements of agriculture, animal health, public health and environment at large. They must also adopt a 'life cycle' pesticide management approach, which addresses various aspects concerning the development, registration, manufacturing, distribution and safe disposal of pesticides and their residues (Copeland, 2007).

These international organizations are further involved in promoting Integrated Pest Management (IPM), as well as Integrated Vector Management (IVM) programs that seek to improve pesticide application and use. They recognize that it's only through alternative vermin control methods that long-lasting and safe solutions can be found for dealing with pests, without having to use toxic chemicals.

To increase public participation in safe pesticide applications, some international organizations also participate in raising awareness through campaigns, poisoning control programs and partnering with other stakeholders involved in sound management of chemical projects (Copeland, 2007).

CHAPTER 9

Handling Pesticides

CONTENTS

9.1 INTRODUCTION

The handling of pesticides is a crucial requirement for all persons who handle pesticides during production storage or even during sale and use of pesticides. It is important for pesticide handlers to know that pesticides are toxic to both animals and human beings and therefore they should take special precautions when handling them during production, transport, storage and during pesticide application. Moreover most pesticides will cause adverse effects in case of accidental or intentional ingestion or even contact with human skins, eyes or inhalation of the same during strain pesticides also may contaminate drinking water soil or even human or animal food and because of all these special care should be taken during transport storage and applications so as to avoid such contamination.

Additionally, people handling pesticides must be properly trained on the proper handling of pesticides to avoid any danger that might be caused by manhandling of the same. It is also important for pesticide producers to take a step to pack and label pesticides appropriately so as to give proper consideration to those expected to handle and use pesticides. Manufacturer's labels should be well stated clearly in a language that will be understood by all who are expected to handle the pesticides. Finally, farmers and other pesticide users should clean the equipment regularly and maintain them to avoid any hazards. They should also read and adhere to the safety warnings on the manufacturer's label additionally they should keep pesticides in its original containers and wear protective gear as recommended by the manufacturer.

9.2 METHODS OF PESTICIDES PACKAGE AND STORAGE

9.2.1 Glass

It includes glass made containers and bottles and it is a protective device such as soluble packages that may also play a role in reducing exposure. Glass packages are made in a way that the pesticide is measured and packaged to dissolve in water without having to open the pack in order to dissolve. During mixing the glass is drained and should be rinsed with volume of water. In addition to protective devices, protective practices can also reduce exposure to pesticides. Their effectiveness is based on the ability of the user to observe hygiene practices. Avoidance of exposure may be achieved by using practices such as spraying downwind. The use of faulty equipment has led to high levels of exposure among applicators.

In addition, farmers often do not refill the container with the same pesticide that was contained in the package before. This may lead to wrongful usage of pesticides, especially in cases where the two pesticides are vastly different and have different uses. Farmers frequently can forget what is in containers, especially if the label pesticide producing countries are partly to blame for this misuse because more often than not they do not inform importing countries of the full dangers of the chemicals. For example, in the study area majority of the farmers applied pesticides without wearing any form of protective clothing and they rarely changed the clothes used during pesticide application which increases the chances of pesticide absorption through the skin. These practices result in over three million people being poisoned annually by pesticides. A commendable effort to discourage the use of developing countries as dumping ground for chemicals was done; importing countries of their health and environmental hazards and why they have been restricted or banned domestically (Cohn, 2011).

Figure 9.1: Pesticides stored in glass.

Source: https://ipm-info.org/pesticides/survey-vegetable-growers-kanchana-buri/pesticides-stored_in-farm-house/

9.2.2 Plastic or Metal Flasks

These containers are on high use especially plastic. Protective devices and practices can play a big role in limiting exposure to pesticides. They are designed to reduce opportunities for human and environmental contamination by chemical agents. Devices may reduce exposure in industrial environments or can be for personal protection. Industrial protection is usually complex requiring the construction of major safety devices for the protection of all workers in the work area. The choice of personal protective devices depends on the degree of hazard. Research has shown that all pesticide workers except those using closed worker systems are exposed to pesticides. However, applicators using plastic or metal flasks knapsacks are the most exposed as compared to ground applicators using other equipment and aerial spray men (Cohn, 2011).

Figure 9.2: Pesticides stored in plastic containers.

Source: https://www.chicagotribune.com/lifestyles/home-and-garden/ct-home-0529-garden-qa-20150527-story.html

The choice of protective devices depends on the work being done and the hazard likely to the associated with it. During the mixing of the concentrate and while applying pesticides in enclosed or when using highly toxic pesticides respirators are needed they are retrieved from it. To be effective, an Aspirator should have an air filter or cartridge. There are various systems available for dust, organic vapors, acid vapors and various other chemicals. However, in the absence of sophisticated equipment, a clean rag tied over the mouth and the nose will reduce exposure. Evaporation, movement in soil, plant uptakc and general metabolism) can be used modeling and predicting environmental impacts in a way that has not been possible in causal analysis using simplistic information such as customary criteria.

9.2.3 Bottles

Protective bottles are devices that have been found particularly useful in reducing pesticide exposure to the skin. The aim is to reduce the surface area that might be contaminated with pesticides and which may be absorbed into the body. To be useful, clothing should not absorb the pesticide as this would lead to increased contact over time. Conversely, clothing does not allow for ventilation and also decreases the evaporation of perspiration. In the search for suitable alternatives, the efficacy of repellent treated fabrics. The workers found that significant protection was afforded to farmers who used cotton denim overalls and laundered them daily when using bottles in various applications in the farms. When a repellent resin was sprayed or spilled onto the clothing, they absorbed less pesticide and were safer to use (Cohn, 2011).

It was recommended that protective clothing should be laundered daily and together with the family washing. Other workers have studied the usefulness of protective clothing in reducing exposure. Nylon-knit gloves and long-sleeved shirts minimize exposure of cut bottles or spill of the pesticides. Gloves can prevent 90-95% of phasolone residues from reaching the hands of peach orchard workers. Different bottles afford different protection. Furthermore, different chemical agents behave differently depending on the fabric used. The choice of a proper fabric for each pesticide is important. One of the factors that affect the penetrability of clothing is the formulation of the pesticide. The Environmental Protection Agency evaluated the issues and problems of protective pesticides in the bottles. They resolved that factors such as permeability of clothing and the penetrability of pesticides through various treated and untreated materials should be involved (Cohn, 2011).

9.2.4 Drums

They are common for large volumes especially industrial use. Eye protective devices are necessary to protect the eyes from splashes of toxic or corrosive materials. Contact lenses are not recommended when working because they may trap the pesticide and increase contact to the eye. Face shields are preferred to safety glasses. The latter is considered the best protection for both physical injury and bright light. Other protective devices that are being used today are suits providing a completely closed system. Both field and factory systems are now in use. It used biological monitoring to evaluate the effectiveness of closed systems in reducing exposure. They found that in mixing-loading and application operations, the use of closed systems resulted in reduced depression of acetylcholinesterase activity in blood and reduced levels of alkylphosphates in urine.

The advantages and disadvantages of these systems and the structures that must be in place before they can be widely used have been reviewed. A proper training program for workers who use the equipment is essential. Leaking pumps, valve springs and nozzles may lead to a lot contamination; therefore, maintenance of equipment can help to reduce exposure. The removal of contamination after accidents such as the quick clean-up of spills is important. One cap was introduced in a contest where children were tested on their ability to open it. Among 300 children, those under three years could not open the cap even after being shown how, and few children under eight could open it without a demonstration. The cap was for use in the pharmaceutical industry but the same principle may be applied for pesticides. Proper storage also takes into account the necessary procedures in case of fire outbreak. Pesticides should be stored in tanks surrounded by dikes as in the case of fuel oil (Cohn, 2011).

9.2.5 Plastic Bags or Paper Bags

In many less developed countries, pesticide plastic bags and paper bags are used for packaging and storing many products, including pesticides. However, pesticide plastic bags and paper bags should not be reused unless for refilling with the same pesticide. Methods for disposal of farm unwanted pesticides and containers are available. Plastic containers must be cleaned thoroughly. In certain cases, it is necessary to clean organic solvents and detergents. The use of pesticides around the home requires that special attention be paid to protective practices. Pesticides should not be stored in large amounts. Children and even have been reported to have ingested pesticides due to poor storage or lack of labeling.

In the home, all food and water containers and eating utensils should be protected. Farm animals should also be removed from areas where pesticides are being used or stored. Old pesticides may pose a hazard in the home. Past studies have proposed a scheme to encourage the disposal of old pesticides around the home. These toxic roundups' can be carried out regularly to clear homes of toxic wastes. Health departments may organize this type of exercise in order to rid homes of unusable pesticides, old paints and any other toxic materials. In different countries, the Pest Control Products Acts regulates the importation, exportation, Manufacture, distribution and use of products for the control of pests. There are also regulations on licensing of premises, imports and exports, registration and labeling, advertising. There is a possibility that safer pesticides may be developed as new technologies develop. An understanding of the mechanism of action of pesticides and especially the specific biochemical site involved in pesticide action is likely to lead to safer usage (FAO/WHO, 2019).

9.3 ISOLATION OF PESTICIDE STORAGE FACILITIES

Homeowners must be made aware of proper storage practices to avoid human harm through exposure by accident or otherwise especially children secure storage facility to ensure fewer cases of theft or vandalism or potential use by unauthorized persons. The storage area should, therefore, be established and kept to a minimum so as to avoid all the above. Discourage storage of large quantities of pesticides. Thus the space of storage should be large enough to accommodate newly-purchased pesticides open the containers and also unused or unwanted pesticides whichever it is possible or recycle. It is important also to ensure the storage facility is away from a man and animal habitats and be a separate structure only dedicated to pesticides storage. For those who cannot build separate storage, a precise area in a building should be designated for pesticide storage. The same as related sites should be safe from unauthorized persons for this reason the site should be always locked with legible signs on doors and windows to warn people on what is stored there. Finally, it is also important to prevent the site from flooding, wind, drainage and sunlight which may affect it suitability for pesticide storage (FAO/WHO, 2019).

9.4 INDOOR STORAGE OF PESTICIDES

Once received from the distributor's pesticides must be stored and the area secured immediately. This is to discourage theft, pets and other unauthorized persons. When storing pesticides on shelves always place the liquid formulations on the lower shelves and the dry formulations on the upper shelves. This is basically to avoid contamination of dry formulation by liquid leaks. Moreover, keeping liquid containers on the lower shelves and reduces the risk of spills if knocked off the shelf. Additionally, to prevent contamination or accidental use, always store pesticides in their original containers with the labels. Do not try to store pesticides in a different container because it's the violation of law (FAO/WHO, 2019).

Daily check the pesticide containers regularly for any leaks or spills or other potential problems. Transfer only the contents of leaking containers to another similar container with the same label to avoid confusion. If not sure about the safety of the leaking container, please contact the manufacturer for directions. To prepare for pesticide application removes the pesticide container from storage facility and takes it to the mixing area or an open field. This is meant to ensure proper ventilation to the safety of those applying the chemicals to the sprayer. Finally keep track of the shelf life of the pesticides; some have short shelf life and cannot stay for longer. For this reason, purchase only needed amounts of pesticides to reduce off-season storage. It is also advisable to remove all the pesticide containers whenever need arise. Always Mark containers and their date of purchase for storage and indicate whether they are still usable to avoid any confusion (FAO/WHO, 2019).

9.5 SECURITY OF THE STORAGE FACILITY

The pesticide storage facility is a very important building or structure which needs serious security programs to keep it safe from theft cases and access unauthorized personnel, livestock. The best and the first way to ensure security is to build a secure facility which is theft- proof and cannot be easily vandalized with the use of material such as bricks and steel windows and doors to enhance security. For extra security measures, put up a strong perimeter fence around the facility to enhance security. Always lock the gate at all times to keep off trespassers or unauthorized personnel. Moreover, ensure the facility has security lighting system at night loaded with a backup generator to avoid any power outage.

Additionally, install an alarm system on the facility to alert in case of breaking by unauthorized persons or wild animals. It's also more important to have a closed-circuit television to assist the owner check the security of the facility. Assisted by the alarm system and CCTV will give you an up-to-date live feed of the facility in case of insecurity. Finally, protect the facility from causing any damage to man for his environment. To avoid this, storage facility containing pesticides should be segregated away from food medicine milk and other edible products to ensure less environmental hazards to man and animal habitats (FAO/WHO, 2019).

9.6 VENTILATION OF STORAGE FACILITY

A good pesticide storage facility must have proper ventilation which meets the requirements and conditions as required by law. The need for proper ventilation in a storage facility will ensure that fumes do not build up in the pesticide storage room. Providing proper air exchange will eliminate unsafe chemical fumes. Secondly, a good ventilation system is good for the storage facility so as to keep of pollutants, bacteria and moisture. The storage facility should have conducive and fresh air circulation in every part for man and the pesticide personnel. Thirdly, well-ventilated room for pesticide storage is a productive workplace thus reduces temperature buildup in hot areas that may harm pesticides.

Proper ventilation is also required to help reduce condensation which may lead to molds consequently leading to health hazards to those who access the storage facility. Pesticide stores also need proper ventilation system to reduce back drafting which is normally caused by pressure imbalances. This situation may cause pesticide bottles and containers to crack especially those made of glass or plastic. Properly built pesticide stores should put this into consideration to avoid fluctuations in pressure. Finally, a pesticide store master well ventilation to avoid the concentration of volatile organic compounds this organic compounds on high concentrations may turn toxic at high levels and endangers the pesticides in the storage facility due to its high concentration (FAO/WHO, 2019).

9.7 LIGHTING IN PESTICIDE STORAGE FACILITY

When working with a pesticide facility it necessary to have a durable and efficient lighting system in each part of the storage facility. The facility should be well lit because those using the building must be able to see properly. Proper lighting helps handlers notice rising problems early enough like container is leaking or eroding and their expiry dates and thus safety precautions taken immediately. The pesticides lighting system should guarantee visibility and safety. There are a wide variety of explosion-proof lighting options for storage facilities. However, available lighting systems should be based on the building size and mounting height of the light fixture.

Light features available include but not limited to the following:

- Standard pendant light fixture which extends 16 feet down from the ceiling. The facility can be raised to match the light fixture or can be located to corners of the walls.
- Linear type fluorescent fixture which is mounted near the center of the ceiling and positioned to give fire suppression components.
- Optional emergency lighting is available in 6 volt battery backup for lighting the path. Emergency lights are designed for wet locations. The lighting is strong to endure weather changes. The emergency lighting system includes a compact assembled system view through windows that allows easy monitoring ultraviolet light and a standard time delay.
- Explosion-proof lighting is made with sealed heavy-duty housing of glass globe and a protective metal guard. Corrosion proof prevents sparks within the facility and reducing the risk of igniting dust gases or vapors surrounding the area.
- Non-explosion proof exterior lighting is used to illuminate the storage facility. The lights can be used continually in wet or harsh climates.

9.8 IDENTIFICATION OF PESTICIDE STORAGE FACILITY

The pesticide storage facilities should be constructed and maintained keeping safety and environmental protection as a priority. The things considered should be building site, storage area, drainage, heat ventilation, labels and contamination. The building site of the pesticide storage facility should be away from sensitive areas like homes playing grounds and should be situated where it cannot endanger day-to-day activities of the living i.e. away from rivers because it can cause water pollution. The storage area should be properly designed to keep the pesticides in good condition which will also keep the environment safe. The storage area should be built away from other buildings for exclusively pesticide storage. The surface the storage facility should be constructed of metal with concrete floors for easy cleaning in case of accidents or spills. The area should have warning signs the signs in an understandable language (FAO/WHO, 2019).

The drainage system of the facility should be built to collect any runoff water. Pesticides which may be in tanks, spill on floors must be controlled to prevent damage to surface water. The collected water should be treated as a dangerous hazard and disposed of immediately. To avoid contamination, the personal side should be stored separately from each other because some can readily contaminate others to not keep food seeds and seeds for fertilizers in the same room with pesticides.

9.9 PESTICIDES SPILLS AND CONTAINMENT

9.9.1 Identification of Pesticides involved in Spill Incidents

Proper storage of pesticides involves regular checks to identify any pesticides that may be involved in spills or leakage. Regular inspections of pesticide containers for cracks spillage splits or corrosion will ensure save from spill. After discovering leaking containers put on protective gear and take appropriate action. Furthermore, if the leaking container is already involved in spillage of pesticides prevent the leakage from spreading and start the cleanup process. However, be extra careful if the spilled pesticide is from an aerosol container or a fumigant cylinder that contains pesticides under pressure (Whithaus, 2016).

After proper identification of pesticides involved in the spill incidents and depending on the situation it is important to consider the following guidelines;

- Apply the pesticide immediately take consideration on the rate allowed by the label
- Let the damaged container and all its contents into a larger suitable container
- Transfer the pesticides to a similar container that has an original label same as the initial pesticide container and ensure the pesticide formulation is the same.
- If you have another partially full container of the same product with almost the same shelf life, transfer the contents there and ensure the container has a lead and the label intact.
- Finally, remember to triple rinse all the leaking containers before declaring them for disposal so as to reduce further contamination to the environment.

In conclusion, it is important to understand the main causes of pesticides used so has to have a remedy in place. Leakage of containers in the storage facility or during transportation is mainly caused by poor handling, mechanical damage and rough driving during transport. All these cases are crucial for one to understand the process of spill containment and control (Whithaus, 2016).

9.9.2 Safety and First Aid to Personnel Handling Spills

When a spill occurs, individual safety and spill containment clean up and consider who to call for assistance. One should be able to train all personnel handling pesticides in the working area. In case of an emergency, procedures must be followed during transport and storage. All personnel should be well conversant with the safety and first aid to be taken in an emergency case with immediate concern. All people should be trained in basic evacuation procedures. They should have first aid training and personal protective equipment readily available and maintained. The telephone numbers of medical assistance and poison control center must be posted on the walls. The exposures of the personnel should be minimized. All the sources of ignition should be turned off to prevent explosion. Injured personnel should be moved to a safe location and contaminated residues removed from the victim. The skin exposed to the spill should be thoroughly washed with soap and water to prevent further damage on the victim additional first aid like flashing eyes or undetected part of the body that were contaminated by spills should be administered with clean water if the symptoms persist get medical assistance for the injured persons.

Figure 9.3: Poorly stored pesticides spill.

Source: http://www.fao.org/agriculture/crops/obsolete-pesticides/world-photos/topic-photos/poor-storage/en/

The site should be secured at all times to prevent an unauthorized person from entering the spill area. Make use of a sign and the barrier tape or rope around the affected area. Ensure that you eliminate all sources of ignition in order to prevent fire explosion. Get help from police to set perimeter around the spill area. Always ensure you wear appropriate protective gear to be on the safe side so as to protect yourself. Leaking containers should be contained always to ensure you prevent pesticides from entering into waterways. Block any entrance to waterways by using absorbent material or racks. Ensure that these spills are covered to stop it from spreading to the other areas leaking containers should be handled with care and separated from others.

9.9.3 Care of the Injured

The most important measure in spill control is tending to the injured or contaminated personnel. Obtain emergency medical attention for such injured personnel as soon as possible (Whithaus, 2016). The following procedures will provide you with a general guide to tend to the injured personnel.

For spills on the skin

- Flush with flowing water for more than 15 minutes and remove all the jewelry for clothes to ensure all residues have been cleared.
- Wash with soap and water particularly warm water if not visible burns are involved.
- Avoid using creams or lotions or selves and less needed.
- Seek medical attention for any chemical bonds

For splashes into the eyes

- Flush immediately with tepid potable water for more than 15 minutes.
- Hold the eyelids away and ask the injured to move the eyeballs up and down and wash thoroughly.
- Seek medical attention immediately

For cuts

- Put on drops in case of contact with blood.
- Flush out with water to flush out chemical contaminants.
- Apply a bandage on the wound and report the injury to their

physician immediately.

- Call emergency personnel in case of a serious injury while providing first aid.

For burns caused by heat

- Call emergency personnel
- Flush with tepid water and apply a moist dresser and a bandage loosely
- Seek medical attention

If the victim is unconscious

- Call emergency personnel
- If safe to enter the area, Place the victim on his or her back and cover with a blanket
- Clear the area from any broken classes and chemical spills
- If the victim starts vomiting turn his or her head so that the stomach contents cannot be aspirated into the lungs
- Stay with the victim until medical assistance arrives for assistance

9.9.4 Spills Site Security

Pesticide security is a very crucial step in pesticide spill control and containment. To be safe, all personnel entering spill site must have the right protective gear.

The following steps should be followed in securing the spill site

- Dust and vapor should be prevented from spreading especially if the pesticide can produce airborne dust or is volatile this is mainly done by closing the storage facility doors and windows to prevent the spread of dust and vapor.
- Controlling the spread of liquid pesticides by making a deck around the outside edges of the spills. Immediately use an absorbent material such as a spill pillow.
- Collect and containing spill cleanup residues. Place all spill residues into a plastic package before transportation for disposal.
- In case of a large gathering outside the pesticide spill area, evacuate the area and declare it a danger zone to avoid trespassing by unauthorized personnel. Use a warning sign to keep off unauthorized personnel.
- Disposal of all waste is the next step. Separate clean up material

of pesticides spill from other trash. Call a health and safety officer for guidance then pack the waste appropriately awaiting disposal.

- Remove all sources that may cause ignition especially flammable liquids by use of pillows only when dealing with flammable pesticides; also switch off all the switches that may ignite fire or cause explosion.
- Finally, decontaminate the spill site and all the other equipment. Ventilate the area by opening windows and doors or even better use a fan. Use conventional cleaning products with a sponge to provide proper decontamination.

9.10 CONTAINMENT AND CONTROL OF SPILLS

All persons using or handling pesticides must always have a back-up plan on dealing with pesticide spills. All people dealing with pesticides must be trained in the procedure and spill management.

- The first step in spills management is to identify the sources of the spill and stopping in to avoid further spillage. Any leaking pesticide container or pipe can be controlled by sealing the leak.
- The second step in containment and controlling of pesticide spillage is to secure the spillage area. Keep away humans and livestock more than 30 feet away. Use protective wear to avoid contact with the spillage fumes for liquid formulation.
- In cases of large spills, call for emergency help from health safety officials. This is to ensure the spillage is contained before it contaminates the environment. Trained personnel with proper protective gear will take care of large spills effectively.
- To contain and control spills, stop the spills using damp soil to prevent the spills from reaching water sources. Use absorbent material for all the liquids as a remedy.
- When handling dry spills cover the same with a plastic tarp. Dry Pesticides will be more controllable because they don't spread faster while doing so avoid the dusting the area because you will generate the dust fumes which may be carried by air.
- The final step in pesticide spill control is to decontaminate all material and personnel involved in spill containment. Wash with soap all body parts such as ears, eyes and nose. Discard all

contaminated leather clothing and boots in case of accidental inhalation or ingestion seek medical attention immediately.

9.11 DISPOSAL OF PESTICIDES AND PESTICIDE CONTAINERS

Leftover pesticides and containers must be disposed of safely. Expired pesticides and their containers must be disposed in a specially dug hole. Most pesticides are toxic and must not be disposed in places that may contaminate drinking water, fish ponds or rivers. The dug hole should be hundred meters away from streams, wells and human settlements (Whithaus, 2016). Containers should be buried without the chemicals after pouring runoff water to rinse them thoroughly and the hole closed as soon as possible. However, for clean plastic containers and cardboard, they should be disposed differently by burning far away from houses and animal habitats.

Thirdly, pesticide residues from non-refillable liquid containers should be removed. This is the cheapest and most economical process which helps a lot in protecting both man and animal habitats. It is done by the use of detergent to rinse containers to avoid exposure to pesticides to people, livestock, wildlife and the general environment. Pesticides bags may also be disposed after thorough shaking of the same to remove all possible solid pesticide formulation. Thereafter bags should be cut in all sides and opened completely to empty its contents. Dispose the empty bags in pit latrines and landfill if the local law allows that. Do not store the bags for future use indoors in the places of residence of man or livestock.

Pesticide waste may also be disposed of by using it to apply on the target animal or plant or structure but be careful not to exceed the required rate. This is only applicable when disposing excess or rather small quantities of pesticides. When disposing large amounts of pesticides, always consult a specialist. Agricultural institutions sometimes sponsor programs on the disposal of excess pesticides. For expired pesticides, to avoid accidental poisoning and damage to the environment transport and dispose the large amounts in the safest place and manner (Whithaus, 2016).

Figure 9.4: Disposal of pesticide containers.

Source: https://sdda.sd.gov/ag-services/pesticide-program/pesticide-collection-program/

During the transportation of pesticides for disposal please follow the following guidelines;

- Keep pesticides in their original containers with the labels intact.
- Place containers in a way they will not tilt.
- Place pesticides in a plastic bin to avoid spillage in case of an accident.
- Do not put pesticides in the passenger section of the vehicle.
- Keep pesticides far away from foods be it livestock or human food.
- Drive safely and carefully to the disposal site.

The final step in waste disposal is to dispose of all material and substance used during cleaning or disposal. First dispose all the rinse water properly so as to avoid contamination of ground or surface water. This is avoided by doing thc following;

- Do not discharge rinsing water to the ground ditches or streams.
- Do not dispose of rinsing water to a septic tank.
- Do not use excess rinsing water to avoid a runoff.

- Do not burry drinking water.
- WARNING; for whatever reason do not re-use rinsing water in your next pesticide formulation.

Secondly, the material used to collect and clean spills should also be properly disposed to avoid future contamination. All cleanup related material must then be collected and disposed appropriately. Finally, dispose-all pesticide-contaminated clothing or materials depending on the type of material and the exposure then dispose of it appropriately.

CHAPTER 10

Rules And Treaties That Govern Pesticides

CONTENTS

10.1 INTRODUCTION

A pesticide is any substance or mixture of substances that is intended on use to repel, destroy, or damage any insect, fungi, weeds, bacteria or any other form of plant or animal that is considered a pest. This includes fungicide, insecticide, bactericide, herbicide, desiccants, plant regulators, defoliants and nematocides. Private, commercial and urban pesticide applicators have laws that regulate the use of pesticides. The laws are provided by the EPA the U.S. Environmental Protection Agency.

The U.S. Environmental Protection Agency (EPA) is responsible for the development of pesticide laws on labeling for safety and assurance of health to users, registration, training of the applicators of the pesticide by providing materials to then and ensuring the application of the federal pesticide laws and regulations. Through its Office of Pesticide Program (OPP), EPA applies/uses FIFRA for the management of its operation.

The EPA's goal is to protect humans and environment/plants by regulating pesticides. This is also realized by enforcing the FIFRA but has the option of leaving the enforcement to the state by agreement or even to the American tribes. The states have laws that govern the use and application of the various pesticides and the laws are very strict. The certification is always done by the EPA. Due to the certification of pesticide users by the state agencies, there must be good and clear knowledge by the applicants on the usage of pesticides and the laid down federal laws (Yan, 2017).

10.2 THE FEDERAL INSECTICIDE, FUNGICIDE, AND RODENTICIDE ACT (FIFRA)

The congress passed the Federal Insecticide, Fungicide and Rodenticide Act (FIFRA). This requires registration of all the pesticides in and into the country. Registered pesticides could be used by anyone to serve any purpose of wish, in any way since the act did not dwell on the pesticide use. It is a United States federal law that looks to protect pesticide users, environment and the general consumers. It is regulated by the United States Environmental Protection Agency (EPA) and environmental agencies of different states. Revision on the law has been made to ensure preservation of human health and the protection and care of the environment by:

- Shifting of the burden of proof to the manufacturer of the chemical to ensure the process of registration is strengthened.

- Making sure that all the products that have been banned and the ones that lack registration comply with the requirements
- Including the regulatory framework missing from the previous law

An applicator may not use a pesticide out of the guidelines of the manufacturer. The pesticide can only be applied on plants, animals or surfaces indicated on the label on the product. No excess dosage, high concentration or over usage of the pesticide. Proper usage and disposal is a must for the applicants.

Penalties: Any person who goes against the rules of FIFRA must be subjected to a penalty of not less than 5000 U.S. Dollars on each offense before the agency issues a fine. Any applicator who willingly and with knowledge violates FIFRA regulations shall be fined 25000 U.S. Dollars or less and if not one year in prison.

To ensure safety and good use, pesticides have to undergo various tests. FIFRA came up with a number of rules.

- All pesticides had to be registered and this s done only after data collection to ensure it is good for the intended purpose appropriate dosage and some of its effects. After registration labeling is done to guide the user. If the user ignores instruction, then he/she is reliable for the hazards caused.
- Just a small number/amount of pesticides is available to the public. Most pesticides are dangerous to the public and are to be used by special applicants only. Distribution of the pesticides is checked.
- There are three categories of pesticides: antimicrobials, biopesticides, and conventional pesticides and EPA have a deferent process to review each. The review process can take long (several years) depending on the different pesticides. State registration requirements may be considered after EPA registration.
- State requests for yearly reports of usage from the pesticide users.

Import and export: Notice of arrival is required for all the imports of pesticides into the U.S. via U.S. Customs and Border Protection. If the notice of arrival is not complete, then the product would not make it through the customs. The notice of arrival lists the product identity, quantity, date the goods arrived in the country, and the inspection place within the country.

Other rules:

- The goods must meet the standards set by U.S pesticide law.
- If the pesticide is not in the exception list, then it must be registered by EPA.
- No violation or adulteration
- Good and clear labeling
- The product must have been produced in an establishment that is EPA registered

Pesticides for exports have no registration requirements under certain cases. The cases are as follows:

- The foreign purchaser has to notify the EPA of his/her knowledge of lack of registration of the product and that it cannot be traded on the U.S. land.
- "Not approved for use in the United States" must be labeled on the pesticide.
- Labels must be written in English and the receiving country(ies)
- The pesticide must meet the FIFRA rules and requirements.
- Must adhere to FIFRA requirements pertaining to keeping of records.

Company pesticide products to be registered with the EPA must have the knowledge of pesticide in accordance with the law. According to FIFRA a pesticide is:

- A substance or mixture of substance that its main purpose is to; prevent from invasion, keep away, or kill any pests.
- Any substance or mixture that is intended to be used on the regulation of plants growing in a particular area.

Any applicant will have to assure that pesticide chemicals used in manufacture, pesticide products, or a proposed new application of a registered product will not cause damage or negative effects to human health or the environment. The negative effects are:

- Risk not common to man to man or environment and takes cost socially, economically and to the environment as well as benefits.
- The dietary risk to food.

Applicant must provide data (scientific) of more than 100 nonidentical tests conducted in accordance with the EPA rules and standards to check on cffccts. Thc EPA can approve of a safe pesticide that can remain on foodstuffs. This is after confirming that it is safe and has no adverse effects on human health. Successfully registered pesticides must conform to approved uses and conditions of usage, which is indicated in the label by registrant (Yan, 2017).

Registration of pesticide: Registration is to ensure health and safety standards are met requirements for labeling, and regulation and moderation of risks. The older pesticide should cause no harm to children, infants and sensitive persons.

Certain products do not meet the requirements to be termed a pesticide. These products include;

- Deodorizers, bleachers, and cleaning agents.
- The products that are intended to attract pests and are never toxic.
- Products that bar away pest by providing a barrier against their access.
- Products that are aimed to be used in control of fungi, bacteria, viruses and other microorganisms in humans and animals.
- Products intended for the control of internal invertebrate parasite or nematodes in humans or animals.
- Products that are targeted to aid in the growth of just a particular species of plant.

Some other products are blocked from registration by FIFRA even though they are considered to be pesticide products. Either way, FIFRA expects proper labeling and branding of the products for proper guidance, safe shipment and application.

These products include;

- Treated article or containing a pesticide that is used to protect the article or substance itself if the pesticide is registered for the use.
- Pheromones and pheromone traps. Only the pheromones that are labeled to be used in the pheromones trap and the compounds are the sole ingredients.
- The preservatives for the various biological specimens that is to be used in experiments.
- Vitamin hormones for plants that are used in the horticultural sector for the flowering.

- Products that has food substance in them and used mainly to attract pests.
- Minimum risk pesticides. These are the products that are not considered pesticides.

FIFRA directs EPA that any data that supports the registration application of pesticides should be made public. If EPA registers the pesticides but certain data are protected as trade secrets and other registrants may not use the very same data to help in registration applications for similar pesticides for 10 years.

10.3 THE FEDERAL FOOD, DRUG, AND COSMETIC ACT (FFDCA)

The united states federal food, drug, and cosmetic act is a set of laws by the congress mandating the U.S Food and Drug Administration (FDA) to monitor and ensure safety of food, drugs, medical devices, and cosmetics. The law was authored by Royal S. Copeland, senator from New York who served three terms (Yan, 2017).

The FFDC has 10 chapters

- Title
- Definition
- Disallowed acts and consequences
- Food
- Drug and devices
- Cosmetics
- Authority in general
- Imports into the country and exports
- Tobacco products
- Miscellaneous

Some of the acts prohibited under the law include;

- The refusal to allow entry or inspection
- Wrong branding of any food, drug or tobacco products
- The introduction into the market any good that is adulterated or branded unlawfully

- Refusal to allow copying of any document or record as the law requires

Food coloring: The act mandated for approval and authorization of some food coloring additives to be a must. Some food colorings are good are not subject to certification by the FDA. Just some food, drug and cosmetics are approved for use and application in the United States and many drug and cosmetic colorings approved to be used in drugs and cosmetics alone. Exemptions to the certification are color additives extracted from vegetables, minerals and animals. Additives got from nature adhere to server standards of safety before a go-ahead in usage for food coloring.

Food additives; The FFDCA requires assurance that the food additives are of no harm to users. FDA approves additives and gives directions and regulations on the usage of the additives.

Homeopathic medications: Homeopathic medications are got from substances in the homeopathic pharmacopeia of the U.S., an official drug compendium in the Act.

Bottled water: Is monitored and considered by the FDA as food. There are set standards and expectations on water quality, labeling of the products and the processing.

Cosmetics: These are items that humans use/apply on their bodies by spraying, rubbing or taken into the human system to change one's outlook, enhance beauty or increase one's attractiveness. Cosmetics that lack proper labeling are removed from the market and do not allow adulterated or misbranded cosmetics. All these are met by supervising and overlooking the manufacture of these cosmetics (Yan, 2017).

Medical devices: All substances that are to be used as a medicine should be classified under the following classes:

- Class I: the products that do not require inspection or a go-ahead in production but must adhere to the rules. An example is dental floss.
- Class II: products cleared using the 510(k) process. Example; hearing assistance devices.
- Class III: Products that pass the test of premarket checking. These are devices that help humans carry on life or are inbuilt into the body. The most common is the Automated External Defibrillator. The products that do not meet the standards of class III are considered class II.

The section of the FFDC act entails that the companies that are to engage in the manufacture of these products, must report and register with FDA, at least 90 days before their plans to introduce a medical product into the market. This is the Premarket Notification. This enables the FDA to group the device in one of the classes and if it's new then proper identification is done to identify the product best.

Any device that makes its way into the market via 510(K) is not approved by the FDA. Either way, it can be sold or marketed in the U.S. they are the "cleared" devices.

Premarket approval: Unlike the 510 (K) pathway, the manufacturer must receive authorization to market the device by writing to the FDA and given a go -ahead. The process of manufacturing and components of the products is embedded in the premarket approval application. This is where study and examination are also taken to ensure that the device is safe for the intended purpose for which it was manufactured.

Amending statuses: Subsequent laws that amended provision of the federal food, drug, and cosmetic act of 1938:

- Food, Drug, and Cosmetic Act Amendments (1951)-Provided definitions and prescriptions for medications. A letter must be taken under the supervision of a doctor.
- Food, Drug, and Cosmetic Act Amendments (1962)-Requires manufactures to show proof of the drug's safety and realization of its intended purpose without posing harm to the user before the federal government's approval. Information on the side effects is also required and this should be provided most so when advertising.
- Food Quality Protection Act (1996)- Amendments provide for health -based standards to govern use of pesticides in food cultivation.

10.4 THE FOOD QUALITY PROTECTION ACT 1996(FQPA)

The act was passed by congress. It was to ensure food safety and pesticide safety. The FQPA transformed the basis of pesticide regulation from risk analysis to health overview. The law provided protection and security for infants and children. The FQPA was supported by the chemical industry, farmer groups, environmentalists, consumers and public health personnel (Reigart, 2009).

10.4.1 Common Mechanism of Toxicity

The EPA is mandated by the Food Quality Protection Act 1996 to check on the substances that cause harm by a mechanism of being toxic. The FQPA came up with ways in which EPA would look into the application of the pesticides and make changes to the Federal Insecticide, Fungicide, and Rodenticide Act and the Federal Food Drug and Cosmetic Act. Some of the expectations in the act target health and set regulations for pesticides used in foods, grant children and infants protection, gave guidelines for the production of good and healthy pesticides and ensure registration of the pesticides.

10.4.2 The FQPA Requirements

Quality standards for foodstuffs and commodity: The FQPA came up with standards that apply to all food substances to ensure safety and causes no harm. The EPA must look into the risk the pesticide causes to children and infants. When checking on the safety of the pesticides, EPA considers the risk at large and the exposure to the pesticides in different ways. EPA is required to establish new policies (Scientific) in order to complete this mission.

Cumulative risk assessment: The EPA developed ways in which they use to test is some chemicals have the same effects. The pesticide addiction is tested. The assessment does not cover all the environmental risk available from the use of the pesticide and this causes some danger to the environment.

Reduced risk pesticide: To be taken as a reduced risk pesticide, it must prove low hazards on human health, low risks, non-toxic and must not affect or damage the groundwater.

Minor uses: The PQPA requires EPA to give special treatment to the products that cover less than 300,000 acres of land and has a low economic impact.

Public health pesticides: The FQPA requires EPA to come up with a number of pests that are important to human health and living and give special treatment to the pesticides that the public uses for health purposes. The EPA provides education and guidelines for safe and good use of pesticides that help control the pest that has been approved to be good for human health. This is also by providing fees on the pesticides that are used to control pest that is advantageous to human health (Reigart, 2009).

Antimicrobial reform: The FQPA disallows other antimicrobial products from the container provisions in the Federal Insecticide, Fungicide and Rodenticide act. The FQPA mandates the monitoring and study of the food consumption behavior amounts to the U.S. population. It also mandated the promotion of integrated pest management solutions. The passing of FQPA saw the ban and placed a limit on the usage of pesticides by the EPA.

Emphasis on children: There was a lot of emphasis on children in the FQPA since the children are considered young and little so the pesticides they take into their bodies have a great effect on them. The EPA ensured the assurance of any pesticide product that the pesticide causes no harm to the children.

Clinical studies: Passing of FQPA brought back the use of clinical studies for pesticide tolerance that was dying. Pesticide manufacturers could limit the errors that were realized when testing using animals by using humans in the test. The manufacturers argued that they must be given a go- ahead to use humans in the test for the pesticides to determine no harm in human health.

The National Academy of Science supported the use of humans in the test of the effects of pesticides on human health in 2004. They argued that this test benefits the community and enables for the determination of the effects of the pesticides hence safety measures applied. The EPA adopted regulations that do not allow the use of expectant women or children in conducting the study/tests. It provided for the maintenance of ethical issues by coming up with the Human Studies Review Board to monitor the use of humans in the study (Reigart, 2009).

The FQPA, Title II, Subtitle A, addresses minor use of pesticides. In support of minor use and regulations, the FQPA will;

- Extend time period allowed for submission of data from the chemistry test and residue.
- Authorize EPA to waive data requirements for the use of pesticides in little volumes or for basic use.
- Direct EPA on the processing of minor use registration application easy.
- Extend the period of registration from the initial 90 days to 180 days to allow the transfer of registration by minor users to other

producers.

- Facilitate the registration transfer for effectiveness.
- Coordinate minor use of pesticides by EPA
- Extends the period of safety use for which one may use the safety data.

10.5 PESTICIDE REGISTRATION IMPROVEMENT ACT (PRIA)

Consolidated appropriations act of 2004 came up with a new way to register pesticides: The Pesticide Registration Act (PRIA). It was a new section, 33, of the Federal Insecticide, Fungicide, and Rodenticide Act (FIFRA) provided for registration service fee system to be used for given pesticide registration, amended registration, and associated tolerance actions, which come up with highest level residue for edibles and feed. The implementation and usage of the law were embedded in the hands of the Environmental Protection Agency and were to act within the period documented. Over 80 different stages of registration application were charged a fee (Yan, 2017).

10.5.1 Authorization of Pesticide Registration Improvement Act (PRIA II)

In the month of October first two thousand and seven, PRIA was given the mandate and authority the start of another financial year of two thousand and eight. PRIA II gave authority to the II: fee routine up to 30th September 2012.

Pesticide Registration Improvement Renewal Act II

- Enlarged the number of charges varieties of being registered after applying from ninety to one hundred and forty.
- Advanced payment for firm employee's security services.
- Well laid provision of financial support for grants for partnering and safe usage of pesticide programs.

The grants for partnership

- Five hundred thousand to seven hundred and fifty thousand dollars per annum.

- Should be given under the initiative of the Pesticide Environmental Stewardship Program.
- Is applied alongside funding to help in Innovative Integrated Pest Management Projects that assist in reducing the effects of pesticides.

10.5.2 The Pesticide Registration Improvement Extension Act (PRIA) III

It came into effect on 1st October 2012. Pesticide Registration Improvement Extension Act III brought back into authority the two thousand and seven Pesticide Regulation Improvement Renewal Act (PRIA II) for 5 years, up to the year two thousand and seventeen.

Pesticide Regulation Improvement Extension Act (PRIA III);

- Enlarged the particulars of payment stages of registering and to apply from one hundred and forty to one hundred and eighty-nine.
- Other petty groups join add as covered PRIA groups.
- Come up with important documented points for completed PRIA groups.
- Maintenance of provision of financial support for agricultural plantation employee's security initiatives.
- Providing financial support for partnering grants and secure training initiatives on pesticides.
- Came up with the provision of funds for information technology improvement.
- Carried on the fifty percent and seventy-five percent payments decrease for the small scale business persons.
- Increased withdrawal from wanted of payment for the agencies of state and for registration by applying recognized with IR-4 meeting a given standard.
- Initiated the payments of a reduced fee by some small scale business ventures.
- Initiate good outcome in receiving PRIA application and for well-stated labeling and label scenarios for the time stipulated.

The Preliminary Technical Screen

This gave a go- ahead to the Agency to spot shortcomings in advance during the time of applying and enables the person applying a ten-day time to rectify the shortcoming at the rare of the task of overseeing.

- A forty-five advanced shortcoming screen time frame for implementation with a period to decide of six months or below.
- Ninety-day advanced shortcoming screen period frame for implementation with decision checking period of more than six months.
- Insufficient registrations that are not rectified may be subject to rejection, hence giving space to agency resources for the applications that can continue to give an outcome of regulation.

The screens are established to ensure all the application components are found. This all included screen does not embed the improvement of the submissions worth.

The tidy/clean labels (2-day label review)

Pesticide Registration Improvement Extension (PRIA III) had in it a mandate to implement (Clean) labeling. Earlier on, other labels may have received approval from the EPA that includes several standards. The labeling requirements were laid down on the joining label changes, which led to putting into use the labels even harder. This had its effects on registering applying process given under PRIA III (Yan, 2017).

The Resolution label time

Effective to RD and AD responses alone and to respond to under the Pesticide Registration Improvement Extension Act III. The Environmental Protection Agency gives the person applying a draft accepted label at that time or some period to the Pesticide Registration Improvement Extension Act laid the final day. The person applying:

- Said yes to any agreement in line with draft accepted label.
- Differed to single or multiple agreements and asked for an extra period to come up with a solution for his lack of conformity.
- Pulled out the applying process with no prejudice for the recurrent submission service charge included.

Anybody applying that asks for more period of time to put into agreement the lack of similarities had until thirty days to come to agree with the Agency on the last agreements of the Environmental Protection Agency recognized the label. If common thoughts were realized, the Environmental Protocol Act had 2 working days after handing in of the reviewed label to review and approval of the last label stamped.

Information Technology Improvement

Pesticides Registration Improvement Extension Act III gave financial support for and expected Environmental Protection Agency to:

- Come up with a private follow up plan of inclusion presentation.
- Come up with tracking of position of inclusion with conditions.
- Improve the database that is exposed to a lot of danger.
- Give room to electronically giving and relook of labels and private documents of formula.

10.5.3 The eight Pesticide Registration Improvement Extension Act of the year two thousand and (PRIA IV)

Since the 8th of March, 2019, Pesticide Registration Improvement Renewal Act (PRIA III) for a time span of 5 years, till two thousand and twenty-three. Pesticide Registration Improvement Renewal Act IV:

- Increase the positions of included categories up to two hundred and twelve from one hundred and eighty-nine;
- Includes improved ways for declined dangerous submissions;
- Upholds small financial operator's payment relief and exempting for federal and state agencies.
- Hikes the yearly cost for maintenance to thirty-one million U.S. Dollars from twenty-seven point eight million U.S. Dollars.
- Does away with the cost of maintenance put for information technology growth/advancements.

10.6 STOCKHOLM CONVENTION ON PERSISTENT ORGANIC POLLUTANTS (POPS)

10.6.1 The Acknowledgments to this Meeting

- To understand and take into account that the waste products allowed to be released into water bodies or carried away by air travel into other countries and expose the environment and water bodies great danger. Considering that the less fortunate communities are exposed to great danger since their environment is exposed to toxic pollutants which affect their plants for food and unhealthy life.
- Having in mind the decisions nineteen stroke thirteen C of 7th February 1997 of The Governing Counsel of The United States Environment Program to include a multi-nation decision to take care of the human lives and the ecosystem by coming up with agreements concerning the stoppage of emission of pollutants into the environment.
- Re-visiting the requirements of the Rio Declaration on Environment and Development Agenda twenty-one
- Taking into account that the gathering and the agreements amongst other countries in issues of doing business and the ecosystem are in support.
- Having in mind the initiatives of the Sustainable Development of Small Island Upcoming States, enforced in Barbados on May 6th, 1994.
- Identifying the assist that the multinational non-governmental organizations contribute to the effort to reduce the emissions of toxic and dangerous waste into the ecosystem.
- Reducing and overlooking the manufacture of the pollutants that are organic thereby involving the manufacturers in an effort to reducing the amount of toxic waste that they release into the environment by giving information to the general public and states to the dangerous components of these chemicals.

- Awareness on the need to take steps to avoid further damages of pollutants that is organic at every step of the chain.
- Upholding the documentation sixteen of Rio Declaration on Environment and Development which reads that individual nations should always evolve funds of their own with an aim of helping cub the release and emissions of toxic waste into the ecosystem without touching on the all nation's agreements.
- Informing the nations that have not come up with rules and regulations that concern the manufacture and application of the pesticide laws to develop them.
- Realizing the benefits of coming up with environmentally friendly activities and chemicals.
- All for the support in taking care of human life and the environment from the dangerous threats of continued emission of toxic waste.

The meeting comes up with the following:

- Aim: taking into account the precautions set in the fifteenth principle of the Rio Declaration on Environment and Development, the main aim of the meeting is to secure human life and the ecosystem from pollutants that are organic.

Some of the initiatives to curb or do away with releases from international manufacture and use, every nation shall:

- Discourage or enforce rules that are legal to ensure the elimination of the manufacture of the substances documented in Annex A to meet the requirements of that Annex; and the bringing into the country and distribution to other nations substances in Annex A in line with the requirements.
- Make restrictions in the manufacture and application of the substances in Annex B in agreement with the requirements of that Annex. Every nation shall act to ensure that a substance in the Annex list is brought within the countries boundaries only when it is intended to be disposed of in an environmentally friendly way; and to serve a purpose that is allowed for the Nations under Annex A and B.
- That a substance documented in Annex A or Annex B for which manufacture or application has particular exemption is in use, having in mind any relevant inclusion in international prior

informed consent instrument, is taking out of the countries boundaries to other nations only: to dispose some waste in an environmental friendly way; to a nation that is allowed to use the chemical under Annex A or B; and to a nation that is not part of the meeting/gathering that has a good document to state the intended use of the chemicals. The statement shall be sure to state; protection of the human lives, comply with provisions of the Annex and comply where necessary with the requirements of Annex A or B.

Inclusion of particular exemption: A list is created to spot the nations that have particular exemptions as documented in Annex A and Annex and the list shall be kept and remain in watch of the secretariat and will be accessible by the public. Any nation can become a part of it by writing to the secretariat, to apply for just a single or more exemptions documented in Annex A or Annex B. In case of lack of registration by any Nation for a specific type of exemption, new registration cannot be realized with respect to it (Copeland, 2007).

10.6.2 Steps to take to Try and Reduce or End Emission from Production that is not intended

- Each nation shall take the following measures to avoid total emission with a goal to reduce it or put it to an end altogether. This can be realized by developing a documented plan to curb this amongst the different nations.
- Try to apply new and available measures in the control of the omissions.
- Promote the use of environmentally friendly and non-dangerous chemicals and control measures.
- Initiatives to curb or do away with emission from dumpsites and waste
- The waste piles in dumpsites that contain the chemicals listed in Annex A or B should be done away with or reduced in a manner that is friendly to human life.
- The scheme for putting into action the plans
- Every nation will ensure the plan to put into action the plans of the meeting.

- The nations shall have a cooperation mechanism to the international and regional organizations including the women and any other institution involved with the children in order to enable the putting into action and use of the rules and requirements.

The sharing of information

- Every nation shall encourage or take part in sharing of information that relates to production, release and use of organic pollution agents and finding a way to do away with organic pollutants and giving information that relates to its effects.
- The nations shall share the message they wish to share directly or via the secretariat.

10.7 ROTTERDAM CONVENTION ON PRIOR INFORMATION CONSENT (PIC TREATY)

The United Nations Environment Program (UNEP) and The Food and Agriculture Organization of the United Nations (FAO) together came up with the 1989 Prior Information Consent procedure. It has assisted in ensuring that nations have the knowledge they should be holding on dangerous chemicals to weigh risk and making up their minds and actions on chemical introduction into their countries (Copeland, 2007). This convention was taken into consideration and presented to be signed at the gathering of Plenipotentiaries in Rotterdam on 10th of September nineteen ninety-eight and enforced on the 24th of February 2004.

10.7.1 Acknowledgements to this Meeting

- Knowledge of the dangerous effects on human life and ecosystem from a portion of harmful chemicals and pesticides that are traded internationally
- Putting into an end the unlawful international trading in harmful products
- Taking into consideration the inquiry of some nations for knowledge on transit movements
- Realizing that managing chemicals soundly should be encouraged within nations

- Identifying that financial exchange and ecosystem rules should help in support with the aim of realizing growth in the economy
- Focused on securing human life and health, employees and the product users included, and the ecosystem from hazardous chemicals and pesticides in the trade that occurs internationally.

10.7.2 The Meeting came up with the following Agreements

The goal of the meeting was to encourage sharing of responsibility and networking/joining effort among Nations in the trade that is practiced internationally of a given dangerous chemical so as to protect the human life and ecosystem from damages and encourage and take part in their good use, by assisting knowledge sharing and by giving for a nationwide process of coming up with decisions on the goods they bring into their nations and take out to other nations and passing the decisions to nations.

10.7.3 Overview of the Meeting

- The meeting is relevant to; the disallowed and unwanted chemicals and very dangerous pesticide formulations.
- It does not apply to; edibles (food), waste, hard drugs, food additives, human and animal drugs or chemically made weapons.
- Installation of National Authority
- Every nation will set aside or install single or multiple authorities authorized to perform factions on its behalf administratively as expected by the meeting.
- Each nation shall ensure the installed authorities are equipped with enough recourse to carry out their factions smoothly.
- Channels to unwanted or adversely restricted chemicals
- Every nation will submit a notice (written) to the secretariat just after implementation and putting into practice rules and agreements of the meeting. The Nations that had done the same were not obliged to do so a second time.
- When the secretariat is convinced of the effectiveness of a given chemical to be used to meet the standards of Annex I, it shall move the chemical to the committee in charge of reviewing the chemicals.

10.7.4 Dangerous Pesticide Formulation Practices

Any growing country that is realizing difficulties with given chemicals within its boundaries and wishes to do away with it can write to the secretariat to inform of the same. This is to list several dangerous pesticide formulations in Annex 3. In coming up with a proposal, the country is able to access the services of experts from other member countries to help curb the menace. In case the proposal contains the message expected; the secretariat will pass the information to all parties after summarization. In case the message does not contain the expected requirements then the secretariat will let the proposer know effectively (Copeland, 2007).

10.7.5 Coming up with a list in Annex III

When an agreement has been made to include in the list a chemical has been realized and the supportive document has been passed by the meeting of the Nations, the secretariat then communicates to all nations without delay. For every chemical recommended for listing in Annex 3 by the Chemical Review Committee, it shall come up with a document for the decision.

10.7.6 The Relieve of a chemical from Annex III

When an agreement has been reached to exclude a chemical from Annex III and the decision guidance piece of documentation has been passed through by the meeting to the Nations, the secretariat then passes this information to all the states.

10.7.7 Information Relating to the Importation of Chemicals Listed in Annex III

- Every party shall put into practice meaningful legislative measures to realize the making of decisions in time in accordance with the bringing into the country chemicals that are registered in Annex 3.
- The final agreement should be in company of a description of administrative action in which it is based.
- Every Nation shall make its feedback under this article present to the persons it concerns with its jurisdiction, in line with its administrative actions.

10.7.8 Information Relating to the Exportation of Chemicals in Annex III

Each Nation that is transporting goods out of its boundaries shall:

- Help in giving decisions and providing assistance to the countries that receive the goods/commodities if requested and effectively.
- Come up and put into practice administrative decisions to pass information about the feedback forwarded by the secretariat in line with the tenth paragraph of the tenth article.

10.7.9 Awareness for Goods and Commodities to Nations

In case of where a chemical that is not accepted or unwanted by a Nation is transported to other nations, and then the Country shall produce an export notification to the nation taking the goods into their country. The note shall have in it the agreements in Annex 5. The country that is trading out the commodities shall produce an up to date export notification just immediately after it has included a final action of regulation that leads to major differences concerning the non-acceptance or inclusion of restrictions of that chemical (Copeland, 2007).

10.7.10 Information that is transported alongside the Goods out of a Country

The information labeled and on the datasheet should, as much as possible, be presented in a single or more of the recognized and official languages of the nation that is acquiring the commodities.

10.7.11 The Passing of Knowledge amongst Parties

Nations that exchange knowledge on a particular idea in line with the convention shall secure any confidential information as agreed. Every party shall encourage the passing of information's amongst themselves that are scientific, technical in nature, economical and legal knowledge pertaining to chemicals with the view of the convention.

10.7.12 The Putting into Practice of the Convention

Each Nation will ensure that the members of the public have the necessary required knowledge on how to use and handle the chemicals and what to do in case of accidents and other ways that are secured to human life and ecosystem as documented in Annex III. The Nations agree to work together at personal levels or through organizations that are internationally recognized, to enforce the laws and requirements of the convention at any level. The convention shall come up with rules and regulations to tackle cases of non -compliance and the treatment the non- complying states will be subjected to (Copeland, 2007).

10.8 THE UNITED NATIONS CODEX ALIMENTARIUS COMMISSION

The Codex Alimentarius Commission structures of operation: The commission shall be answerable for giving their proposals to, and shall be inquired from by, both the Food and Agriculture Organization (FAO) and The World Health Organization (WHO) Director-Generals on issues that pertain to the enforcement of the FAO and WHO standards of food initiative. Intended purposes are: securing of the human health and ensuring that fair prices are charged for the trade of foodstuffs, promotion of joint work of all the standards of food work done by both the non-governmental organizations and the global organizations and making corrections to developed standards, appropriately, with a view to encouraging development. The interest of joining the commission is available to all the nations that are part of FAO and WHO which have a great interest in global food standards (Yan, 2017). The Directors- General of FAO or WHO must be notified for membership consideration.

Any State that is interested in the activities of the commission but does not belong to FAO or WHO may write to the Directors- General of either the UN organizations and will have to attend the commission's meetings and gatherings as will be stated. States that do not find themselves in FAO or WHO but belong to the United Nations may receive invitation to attend the commission's meetings as observers as documented. The commission can develop and use rules/guidelines that will be enforced after the FAO and WHO Directors- General approval that they can be adopted by the commission.

Set regulations of procedure of The Codex Alimentarius Commission: Interest to become a member of the Codex Alimentarius Commission is possible to all the nations that are part of WHO/FAO and associates to the two UN commissions. The regional economic integration institutions or organizations members of either WHO or FAO shall be considered for membership by making the Director-General of WHO or FAO of their interest be part of the commission. Every member of the commission shall make communications to the Director-General of WHO or FAO the particulars (names) of the people who those representing them and where necessary its delegation before the beginning of any meeting of the commission.

The statute or regulations to member organizations: No voting or holding of positions in the office in the commission by the member institutions/ organization or any other body. The member institutions are not allowed to vote in any elections in the commission. Member institutions are allowed to take part in issues within its interest in any of the meetings of the commission in which any of its member nations is allowed to take part in. Member institutions shall practice membership entitlements on a partial basis with the member Nation that are commission members of the commission in the parts of their individual competence. Any committee member may request a member institution or member State to give answers to some of the matters that arise which is in line with the member state (Yan, 2017).

Statute on officers; Meetings are to be guided by the chairperson or the vice-chairperson in his absence and carry out the other activities as expected to enable the smooth running of the commission. The vice-chairperson has equal authority as the chairperson when in an acting capacity. Additional rapporteurs may be up for an appointment by the commission from amidst the delegates of the commission.

Statute on the coordinators: A coordinator may be identified and handed responsibilities from among the members of the Commission from any area/ nation enumerated by the commission in accordance with proposals by a majority of the commission member which consist of the area/region.

Statute on executive committee: The chairperson and vice-chairperson shall automatically become the chairperson and vice-chairperson of the executive committee respectively. The executive committee shall be answerable to the Commission.

Statute on sessions: Geographical area of the meetings/sittings shall be decided by the director-general of WHO and FAO after seeking advice where needed, with the agreement of the chosen country. The commission's gatherings shall be held in public unless the commission develops other plans. Notification on date and place of meetings shall be presented to member countries at least 2 months earlier.

Statute on Agenda: The adoption of an agenda shall be considered from the first particulars on the provisional agenda. Any member can request for inclusion of particular items in the provisional agenda by doing that through the director-general of FAO or WHO.

Statute on voting procedures: The commission's decisions shall be determined by the most votes cast in a particular direction unless otherwise. A roll-call vote may be an appeal by any member of the commission, and therefore each member casting shall be captured and recorded.

Statute on observation: Any member of the commission may join as an agent to observe the proceedings of the subsidiary bodies and may hand in memoranda and take part in proceeding without a vote in matters.

Statute on records and reports: The commission's report shall be presented to the director-general of WHO and FAO after every sitting, who shall pass it round to members of the commission, to other Nations and institutions that were represented at the meeting, for their contribution, and upon request to other member countries and other associate members of WHO and FAO.

Statute on subsidiary bodies: The commission may come up with some subsidiary bodies. For example, subsidiary bodies that it considers important to finish its activities in the finalization of draft standards. The commission shall come up with the subsidiary bodies except where otherwise stated or considered by the statute. The geographical area and time of the meetings of the commission shall be passed to the member nation two months earlier.

Statute on elaboration and acceptance of standards: The commission shall try its best to come to an agreement on the inclusion or changes of standards. The decisions to take or change standards may be decided by casting votes only incase such efforts to reach agreement have not succeeded.

Status on budget and expenses: The expenditure estimated shall include the allocation of travel allowances including daily allowances of the Executive committee members from the underdeveloped nations for the goal of participating in gatherings of the Executive committee.

10.9 ENVIRONMENTAL PROTECTION AGENCY (EPA)

The Environmental Protection Agency`s mandate is to secure the human lives and the ecosystem at large. It was formed in order to execute/enforce major pollution control initiatives that were engaged by the EPA workers at the federal level.

More funds are allocated to EPA, in the budget, to enable continuation in works of addressing change in climate and good quality of air in the surrounding, protecting the water bodies and striving towards a sustainable ecosystem for all the American citizens.

The Presidents FY 2016 budget enrolls funds for the U.S. Environmental Protection Agency to secure people's environment, health and lives. This includes:

- Creating a change in the societies across the Nation through on-the- ground efforts, waters in the urban areas, community resource initiators and circuit riders, advance supervision technologies, and technical help for climate change adaption.
- Giving hand to the head of states climate action initiative to help curb carbon pollution from the power plants, cars, by using safe energy sources and cutting waters.
- Enforcing operations in water resources with an aim of making life sustainable, innovative financing and resiliency together with help to small communities via the new center of infrastructure and resilience finance.
- Enhancing the safety and security of chemical entities by increasing assistance to the local communities to enhance the exchange of information, networking, and capacity, releasing the White House Order on increasing chemical facility safety and protection.

The budget also entails funds to drive The Clean Power State Incentive Fund to assist the states that wish to move beyond the Clean Power Plan.

The president's budget affirms that the Environmental Protection Agency upholds its commitment to effective and reliable government by:

- Enhancing the business of environmental protection for states, the regulated community and the general population through electronic enterprise business model, in which EPA undertakes its responsibilities with co-regulators to smoothen the work process, while improving data quality, sharing of information and accountability.
- Building and renovating structures that are used as laboratories for testing and scientific research take place on behalf of the American population.

10.9.1 Mandates

Realizing a positive change within the communities: The Environmental Protection Agency will place a lot of concentration in spending funds to come up with initiatives that are aimed at supporting good society's environmental efforts, not forgetting the people in marginalized areas and less developed communities. EPA`s new initiatives are modeled to come up with and strengthen the adaptive capacity of less serviced communities through, tools, engagement in training activities and enable the availability and ease of reach to data information. With the application of new techniques, EPA will help the societies address the menace of pollution and detect pollution in the atmosphere and water digitally. The EPA detects staff that they employ to work as Community Resource Coordinators, in the agency to help in service delivery to the many members of a community. The EPA will support the Non State partners to re-energize land.

Tackling the change in climate: The budget provides for a lot of funds to enable EPA to realize its efforts to help find a solution to climatic change through settings of guides and enrolling voluntary programs. Clean Power Plan is a top priority for EPA and the climate agenda. The Clean Power Plan is a practical action that is enforced on the states, towns, and businesses across the Nation to find a solution to dangers of climatic change by checking in the air pollution by existing power hubs. The budget includes funds to help state develop their own Clean Power Plan project.

The EPA will work for hand in hand with the National Highway Transportation Safety Administration to see into climatic change by overseeing the transportation sector's largest leaders to the emission of greenhouse gas. The aims are to implement greenhouse gas rules for both light and heavy-duty vehicles thus reducing carbon emission.

Provision of assistance in needs of water infrastructure by the community: The increased challenges to climatic change create more chances for innovation of new ways to safe drinking water. The EPA receives funds that are aimed at enabling the production and assurance of clean drinking water and better treatment of wastewater infrastructure. The EPA will come up with better solutions such as green roofs, wetlands and rain gardens that will bring a solution to clean drinking water and again restore and support the water bodies that are in the state such as rivers and lakes. A new water center will concentrate efforts on agendas such as financial framework for the future public infrastructure, working with the government to spot opportunities of providing finance to the little communities' agriculture training, assistance and financing opportunities to the rural population.

Provision to realize improvement in safe chemical facilities: The EPA ensures application on needed efforts to help advance the safety and protection of chemical facilities and cut the danger of toxic chemicals to company employees and operators, the human population and responders.

The EPA addresses handling and keeping of the chemicals that must be kept safe and responsibly because of the harm/damage they can cause to the human population and the environment at large. Environmental Protection Agency has partnered with other federal counterparts to come up with federal regulatory representatives with an aim to reduce the dangers that are accompanied with the storage and handling of chemicals. The joint concern includes facility owners and operators, regional bodies, private non-governmental organizations, facility employees and the local environmental organizations. This is for better feedback on matters pertaining to the environment that will assist the Environmental Protection Agency improve its operations. This also helps the agency modernize policies and regulations.

Improve environmental protection by setting priorities collaboratively: For the realization of effective environmental security, the EPA joins efforts with other stakeholders such as the government and community partners. The partners share the goal of coming up with integrated environmental entity for the nation.

The budget allocates for funding to enable the smooth operation of the partnerships to realize a common goal.

The Clean Power Plan will see on the extensive work with the government, communities and nations to improve infrastructure, provide assistance, and build capacity to enable the realization of goals.

10.10 UNITED NATIONS LONDON GUIDELINES FOR THE EXCHANGE OF INFORMATION ON THE CHEMICAL IN INTERNATIONAL TRADE

The codes of United Nations guidelines for the exchange of information on chemicals in international trade got amended in 1989. This set of instructions is geared to address the government to assist them increase chemical safety. The guidelines both cover for any import or export chemical. They, therefore, help the government by exchanging of chemical information in international trade. Universal elements and principles are drawn from the bilateral, regional, and global instruments and national regulations have been used to set these guidelines alongside the already gained experience through preparations and implementations. These guidelines are, therefore general. Therefore, they are purposed to enhance smooth chemical management through the exchange of technical, economic, scientific, and legal information (Copeland, 2007).

For the protection of human beings and the environment all over the globe, special amenities have been incorporated in the guidelines regarding chemicals that are banned or severely restricted in international trade. It is upon exporting and the importing countries to co-operate and have joined responsibility. Their co-operation ensures the protection of human beings, and the environment from banned and severely restricted harmful chemicals.

There are general principles that govern these guidelines. The exporting and importing states should ensure that human beings' health and the environment are protected from any possible harmful effect by exchanging information about the chemical in international trade. If it is a pesticide, the exporting state should provide and ensure that the importing state has the proper information about such pesticides. This makes the importing country know how to handle the pesticide, avoiding any possible harm. In case of any damage to human health or the environment from that pesticide, the exporting state and importing state are answerable to the international trade board (Copeland, 2007).

When any state is taking measures towards the protection of human health, animals, plants, and the ecosystem at large, it should ensure that it should not affect international trade in any way. The states, therefore, avoid any incidence where an obstacle can be created in international trade. These obstacles mean that it will affect the other countries which are not involved in that trade. Therefore, the involved states should look for the best way to protect the ecosystem, not affecting international trade.

Besides, the exporting state and importing state are responsible for ensuring effective and appropriate infrastructure in the two nations. This happens by creating and giving strength to regulatory and legislative systems and the other vital techniques improving the running of a chemical and its management. They also create an industrial and pesticide register of toxic chemicals. On the other hand, they should prepare and update manuals, directories, and documents ensuring better usage of facilities for information gathering. The preparation of these manuals, documents, and directories also provide dissemination at the country level and to region level online facility.

Some private-sector bodies are involved in chemicals in international trade, including pesticides. These private-sector bodies should be committed to undertaking self-operation measures, thus meeting the conduct standards set out principles and codes carried in this guideline to ensure safe production and management of chemicals in both domestic and international trade. Besides, they should concede the commitment of their responsibility, alongside the governments of countries exporting and importing chemicals. This ensures the protection of human health and the environment. In particular, businesses and industries are accorded to acknowledge their responsibility by fully participating in the implementation and evaluation of activities in international trade concerning chemicals (Copeland, 2007).

To reduce and avoid risks, involved states and private sector parties should make reasonable efforts. They should be putting in place the right procedures to minimize unpropitious health and environmental effects from chemicals, under both normal operating conditions as well as emergencies. Again, they should develop safer packaging, and ensure interpret and incisive labeling, taking into account the existing international scheme concerning packaging and labeling. Also, they are supposed to take initiatives, to the possible extent of following chemicals to the final consumer, keeping track of any problems that can arise in the actual use of the chemicals, as a requirement for changes in chemical labeling, directives, and packaging (Copeland, 2007).

On the other hand, whenever possible, the industries should venture to lessen the number of harmful chemicals used. This can quickly happen by co-operating with government authorities in events relevant to the prevention of chemical accidents, receptivity, and response, incorporating the establishment of emergency awareness plans and support all international events.co-operating with the government, the industries should then ensure safe management and disposal of chemicals.

As from the codes of this guideline, Chemical producers, formulators and traders are supposed to play a very critical role. Both of them should Co-operate with concerned government mastery of importing countries and comply with their Prior Information Consent decisions. By their co-operation, they should admit that this might be dependent upon the governments of exporting states, outreaching their culpability to transmit to their industries the Prior Informed Consent decisiveness of importing countries under the Prior Informed Consent procedure. Again, their Co-operation with the government bodies ensuring incorporation of the export notification procedures for banned and severely restricted chemicals is paramount (Copeland, 2007).

As by the codes of this guideline, only known are the ones allowed to be produced and commercialized. A chemical is regarded as known when it undergoes testing and assessment process through which is conducted following national laws and regulations or internationally allowed policies and updated where necessary. The testing and evaluation should give the essential basis of an assessment of any risks of thc chcmical to allow proper actions to protect human health and the environment. Therefore, summaries of the reports of such testing and assessment have to be provided to government authorities. When requested, the full reports should be provided to these authorities per applicable national laws and regulations, in cases where such laws and regulations have been in force. Seeking feedback from users on the use and misuse of chemicals helps to identify possible uses and misuses of the chemicals hence ensuring the safety of human health and the environment (Copeland, 2007).

10.11 THE ENDANGERED SPECIES ACT OF 1973

The Endangered Species Act was signed by the United States on March 3, 1973, protects the endangered species of the world. The United States Fish and Wildlife services and the National Marine Fisheries of the National Ocean and Atmosphere administration are the bodies that administer this act. They are therefore responsible for declaring a species endangered. The law thus provides conservation of the ecosystem by set rules which ensured that species like fish, wildlife and plants aren't violated in any way.

This agency allows reporting of any unintended exposure of pesticides and harm from pesticides. This kind of exposure or harm is known as a pesticide incident. Pesticide incident involves human beings, domestic animals, plants and wild animals. For any pesticide to be used, it should be licensed by the environment protection agency. Therefore, reporting any pesticide incident to this agency helps it to know how to engulf the problem. Here now comes in the Endangered Species Act which then protects any endangered species (Reigart, 2009). The endangered species act permits the payment of any individual who provides furnished information which can lead to arrest or conviction of any violation of the endangered species. This endangerment may be due to the use of unlicensed pesticides. When one is caught or reported causing a threat to such species, he or she undergoes the penalties as stated by the act.

Under The Endangered Species Act (ESA), more than 1,300 species are listed as endangered or threatened in the United States. The two United States bodies administering the ESA has the responsibility to record and delist species endangered or threatened. The act gives the authority for any species to be referred endangered or threatened. For a species to be referred to as endangered, it has to decrease by a rate of 50% - 70% in ten years in a geographical area of fewer than 5,000 km^2. The same geographic population decrease up to less than 2500 adults also dictates that the species is endangered. When the species is noted endangered for it to receive the protection of the Endangered Species Act, it should be added to the federal list of the endangered species. A species is referred to as threatened due to a specific species life mechanism and natural or artificial factors which include the weather changes. Besides, species can be intimidated by the pesticides used in the surrounding environment (Reigart, 2009).

Moreover, this act does not allow possession, taking, selling or transportation of unauthorized endangered species. The prohibition protects the endangered species from extinction, making then to keep on existing. These are some of the strategies that protect these endangered species. The secretary lays down some plans to recover the endangerment and threat to such species. Prohibiting the possession, taking, selling and even transporting, therefore, acts as a way to initiate the recovery of the endangered species.

The act also authorizes the possession and conservation of acquired land to conserve the already listed endangered species. These lands are preserved using the land and water funds provided by the United States and the supportive non-governmental organization. Protection of such properties and environment aids in the protection of endangered species since they are put together in the staffed and protected area which is under care. Therefore, the recovery of these endangered species becomes easy and controllable when the remaining number of individuals are all together (Reigart, 2009).

Also, these two bodies evaluate the status of each species every five years. They probably carry this practice to check on the recovered species, the still endangered species and the newly endangered and threatened species. Having a report, they get the go-ahead of how to handle and the way to take to ensure the protection of such species.

Due to the above set rules on the endangerment of some species, the Environment protection agency provides pesticide information to the farmers, household pesticide users and pesticide applicators on the effect of the pesticides in the ecosystem. For example, the improper use of the anticoagulation of prairie dog bait can harm the non-targeted species, therefore, endangering them. The anticoagulation of prairie dog bait can affect other species directly or indirectly. The non-target species are affected by eating this anticoagulation of prairie bait leading to toxic effects or consuming poisoned prey. Due to this effect, there are limitations that are set to ensure the safety of the non-target species. The anticoagulation of prairie dog bait is not allowed for use in all 13 black-footed ferret reintroduction areas. It is also not allowed within the Blackfeet and Crow reservations in Montana, within five counties in southwestern New Mexico to protect five species (Reigart, 2009).

The household pesticide users should follow the following guidelines to ensure the safety of the endangered non-target species; they should store and dispose of pesticides properly. They should follow the instructions from the instruction label about storage and usage. Additionally, they should buy only the amount of pesticide to be used to avoid stockpiling. One should not reuse a used container for it may contain harmful germs to the non-target species.

Farmers should follow the characteristics of the application before applying the pesticide and beware of the geology as well as the relative depth of the soil to avoid leaching pesticides. For bee protection, the farmers should be conscientious when planting the treated seeds because the dust can cause an effect on the bees. Additionally, while applying the pesticides, they should leave an untreated vegetation border between the treated and the wildlife residence.

The administrators of the Environment Protection Agency in cooperation with the secretaries of interior agriculture had to study. The purpose of the study was to provide a reasonable and prudent report on means to implement endangered species pesticide labeling to the congress one year after the enactment of this act as provided by the law which was enacted on October 7, 1988. Due to the report they gave, the endangered species are well protected from the pesticide effects since they are mostly part of the non-targets (Reigart, 2009).

References

1. Bhadriraju Subramanyam & David W. Hagstrum. (2012). Alternatives to Pesticides in Stored-Product IPM. Berlin: Springer Science & Business Media, 29-73.
2. Christos A. Damalas & Ilias G. Eleftherohorinos. (2011). Pesticide Exposure, Safety Issues, and Risk Assessment Indicators. Environmental Research and Public Health, 8(5): 1402–1419.
3. Cohn, J. (2011). Pest Control Worker. London: Marshall Cavendish, 4-19.
4. Copeland, C. (2007). Pesticide Use and Water Quality: Are the Laws Complementary or in Conflict? New York: DIANE Publishing, 1-7.
5. Ettore Capri & Dimitrios Karpouzas. (2011). Pesticide Risk Assessment in Rice Paddies: Theory and Practice. Netherlands: Elsevier, 1-69.
6. FAO/WHO. (2019). Managing pesticides in agriculture and public health: An overview of FAO and WHO guidelines and other resources. London: Food & Agriculture Org, pp.75.
7. Fitzhugh, B. (2012). Pest Control: An Assassin Bug Thriller. New York: Poisoned Pen Press Inc, pp. 320.
8. Heikki M. T. Hokkanen & James M. Lynch. (2003). Biological Control: Benefits and Risks. London: Cambridge University Press, 3-75.
9. Hideo Ohkawa, Hisashi Miyagawa, & Phillip W. Lee. (2007). Pesticide Chemistry: Crop Protection, Public Health, Environmental Safety. New Jersey: John Wiley & Sons, pp. 538.
10. Khan, S. U. (2016). Pesticides in the soil environment. New York: Elsevier, 1-35.

11. Levine, M. J. (2007). Pesticides: A Toxic Time Bomb in Our Midst. New York: Greenwood Publishing Group, 1-65.
12. Matthews, G. (2008). Pesticides: Health, Safety and the Environment. New Jersey: John Wiley & Sons, 1-78.
13. Patrick O'Connor-Marer, Susan Cohen, & Patrick J. Marer. (2006). Pesticide Safety: A Reference Manual for Private Applicators. London: UCANR Publications, 1-106.
14. Philip J. Landrigan & Mary M. Landrigan. (2018). Children and Environmental Toxins: What Everyone Needs to Know. London : Oxford University Press, 1-73.
15. Racke, K. D. (2003). Release of pesticides into the environment and initial concentrations in soil, water, and plants. Pure and Applied Chemistry, 75(11-12), 1905-1916.
16. Reigart, J. R. (2009). Recognition and Management of Pesticide Poisonings (5th Ed.). New York: DIANE Publishing, 33-125.
17. Soloneski, S. (2014). Pesticides: Toxic Aspects. Berlin: BoD – Books on Demand, 29-81.
18. W.H. Hallenbeck & K.M. Cunningham-Burns. (2012). Pesticides and Human Health. Berlin: Springer Science & Business Media, 1-46.
19. Whithaus, S. M. (2016). The Safe and Effective Use of Pesticides, 3rd Edition. London: UCANR Publications, pp.386.
20. WHO. (2010). WHO Recommended Classification of Pesticides by Hazard and Guidelines to Classification 2009. London: World Health Organization, 1-65.
21. Yan, H. (2017). Pesticide Law and Compliance Decision Making: A Case Study of Chinese Farmers. Berlin: Springer, pp. 262.

INDEX

U

V

W